Did you know?

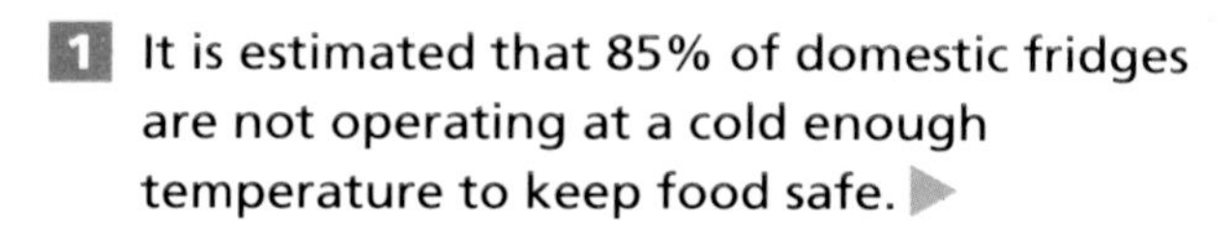

1 It is estimated that 85% of domestic fridges are not operating at a cold enough temperature to keep food safe. ▶

2 An egg-case produced by a German Cockroach may contain as many as 36 cockroach eggs.

3 Food handlers can personally be fined up to £2,000 for smoking in rooms where there is open food. ◀

4 In a year one pair of house mice and their offspring could produce a total of 2,000 young. ▶

5 One bacterium (germ) can multiply to one hundred million in only nine hours.

6 One hundred million bacteria would be no bigger than a pinhead.

7 Rats spend 25 % of their time grooming themselves. ▶

8 There are 3 million people within the food industry from plough to plate.

CHAPTER 1

What is food hygiene?

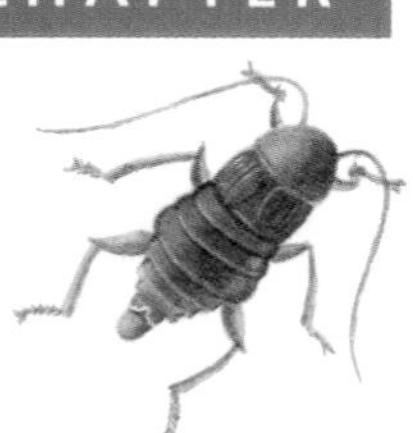

We all need food to live and it is important that this food doesn't make us ill. To prepare and supply food which is safe needs careful food hygiene. This avoids food poisoning, wastage of food, loss of business and legal action against you or your workplace.

WHAT IS FOOD HYGIENE?

Keeping workplaces, staff and equipment clean is an important part of **food hygiene**. If you work in a clean area:

- it reduces the risk of producing harmful food
- it prevents infestations by pests like flies, mice, etc
- it is more attractive to customers.

Food firm fined after woman swallows plastic

USEWIFE needed an emergency operation to ve a piece of plastic from her throat after eating -cooked meal, a court heard yesterday. ... in Mrs Pauline Weston's ... but ... wer ... some

Dirty shop is fined £13,50

THE owner of one of Colchester's oldest cake shops, has been fined £13 ... hygiene laws.

Colchester magistrates im... year-old Patricia Drake, w... counts of contravening several food hygiene regulations, some l...

Landlord is fined £900 for pub dirt

A CAMBRIDGE landlord has been fined £900 for failing to keep his pub clean.

Restaurant fined for having dirty kitchen

Salmonella hotel fined

LEICESTER: Two hoteliers were fined £3,000 ea after 66 guests who ate turkey at a wedding receptio went down with salmonella poisoning. John Kento and Gerald Smit... ...itted breaking food hygien... Brooke... ...l departm... Husba...

Missing tiles cost brewery £200 fine

Cockroach was found in meal

A CHINESE take-away in Edmonton has been forced to shut its doors after a

Where would you rather work? ▼ ▶

An equally important part of food hygiene is the way in which food is handled and stored.
Poor practices in the handling and storage of food could lead to a food poisoning outbreak, even in the cleanest of workplaces.

AVOIDING FOOD POISONING

Poor food hygiene leads to illness and even death.

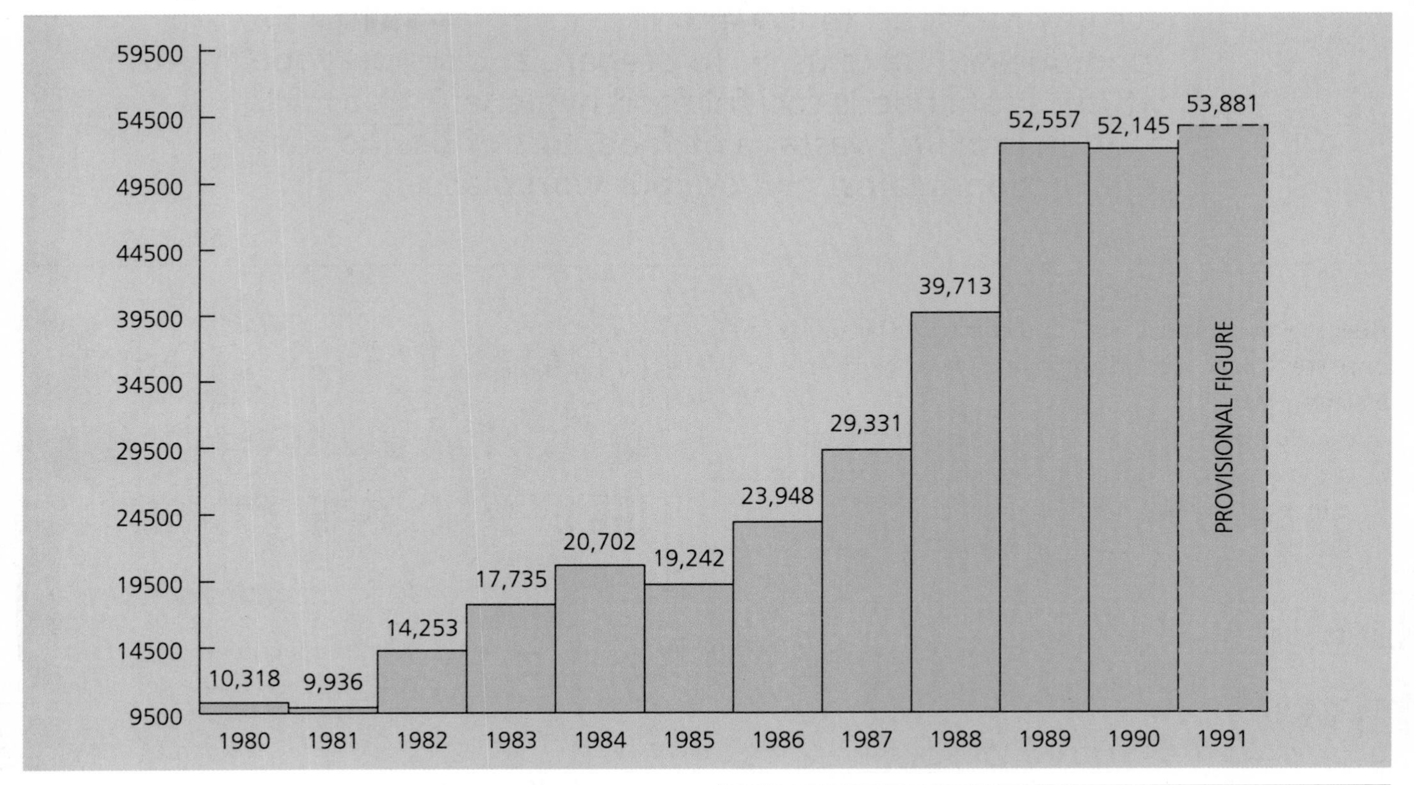

▲ This bar graph shows the number of reported and otherwise notified cases of food poisoning in the years 1980–91. The figures are supplied by the Office of Population, Census and Surveys.

There are up to 40 deaths each year due to food poisoning. The people who die are often those in a **high-risk group**. These are:

- the very young
- the elderly
- people who are already ill.

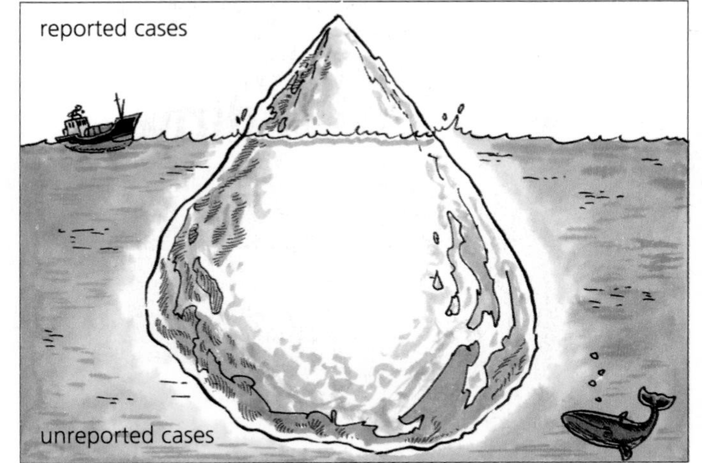

▲ It is estimated that only one in ten people report their food poisoning.

WHY LEARN ABOUT FOOD HYGIENE?

Everyone should know about food hygiene. Nobody wants to make people ill. You should receive training so that the food you produce is as safe as possible. Good habits will protect food against being contaminated or infected by germs. Food should be cooked properly to stop it making people ill.

As well as illness, poor hygiene can lead to:

- food **contamination** by germs and other objects
- food wastage
- infestation by pests, eg flies, mice
- loss of working days through illness
- low production and efficiency
- loss of customers and profit
- legal action against you or the firm you work for.

KEY WORDS

Food hygiene – the good practices which lead to clean workplaces and the safe production of food.
High-risk groups – people who are likely to be seriously ill as a result of food poisoning. Extra care should be taken when providing food for people at high risk.
Contamination – food is contaminated if it contains something which shouldn't be there, eg a stone, chemicals, bacteria (germs).

The law and food hygiene

The Government passed the Food Safety Act to make sure that all food on sale is safe to eat. This affects everyone working in the food industry – in producing, processing, storing, distributing and selling food. The law came into force on January 1st 1991. It is there to protect everyone. Enforcement officers make sure that people obey the Act.

THE LAW

The Food Safety Act 1990 gives ministers power to make regulations. Some of these may be related to food hygiene.

The Act gives enforcement officers power to:

- enter food premises to investigate possible offences

- inspect food to see if it is safe

- take suspect food away from the food premises and have it condemned by a Justice of the Peace if it is unsafe.

You must give enforcement officers information and help when they ask for it – you can be fined if you don't.

If the Food Safety Act is broken it can mean:

- the closure of the business

- a fine of up to £20,000 for each offence

- a prison sentence of up to two years

- compensation for customers who have been affected by the food.

Food hygiene and business

Poor food hygiene affects business. Nobody will want to buy food or eat in a place with a bad reputation. People don't want to work in these places, so there is a high turnover of staff.

They gave me a dirty knife and fork!

I won't go there again – there was a beetle in my rice

A bad reputation can lead to:

- less business
- low profits
- possible redundancies.

Food hygiene means keeping premises, staff and equipment clean and handling and storing food safely. Food should be prepared as safely as possible to lower the risk of illness. Good practices should be taught and followed.

Good food hygiene prevents disease and injury. Poor food hygiene can lead to outbreaks of food poisoning. There were over 53,000 notified cases in 1991 in the UK.

Food poisoning causes serious illness, particularly among **high-risk groups** of people. These are the very young, the elderly, and people who are already ill.

As well as causing injury or disease, poor hygiene leads to:

- food contamination
- food wastage
- infestations by pests
- loss of working days
- low efficiency and production
- loss of customers and profits
- legal action.

The Food Safety Act 1990 is now in force. It covers all food premises AND all food workers, including voluntary workers.

Fines, imprisonment and the closure of businesses are the penalties for breaking the law. You may also be sued by the customer.

Hygienic premises are better, happier places in which to work.

But how can food make you ill? Now read chapter 2 ...

2

CHAPTER

Food and disease

There are several ways in which food can make us ill. Food poisoning and diseases carried by food are caused by microbes, usually bacteria. You can also become ill or injured if there are unwanted objects, poisons and chemicals in the food.

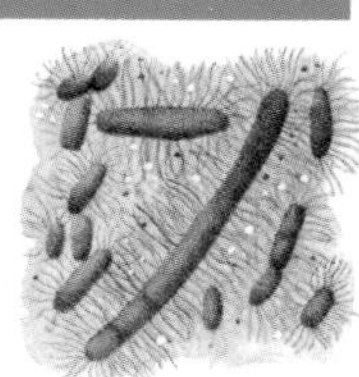

WHAT IS FOOD POISONING?

Food poisoning is an illness you get by eating **contaminated** food. Food is contaminated if there is something in it which shouldn't be there.

Symptoms of food poisoning are:
- abdominal pain – stomach-ache
- diarrhoea – 'the runs'
- vomiting – being sick
- nausea – the feeling of sickness
- fever – a raised temperature.

You may not have all these symptoms. The symptoms can vary depending on what causes the food poisoning.

The symptoms of food poisoning usually start between one and 36 hours after eating the contaminated food and they can last for days.

Food poisoning can be caused by:
- chemicals and metals
- bacteria and other microbes (viruses, moulds)
- poisonous plants (eg toadstools, berries).

CHEMICAL AND METAL CONTAMINATION

▲ Some of these chemicals and metals are poisonous to us and cause food poisoning symptoms. Metals like lead and mercury stay in our bodies for a long time and some will make people very ill.

How to detect chemicals

The food may taste or smell funny, but most chemicals can only be detected by testing the food in a laboratory.

BACTERIAL CONTAMINATION

Food can also be contaminated by bacteria.
This can cause food poisoning or a food-borne disease.

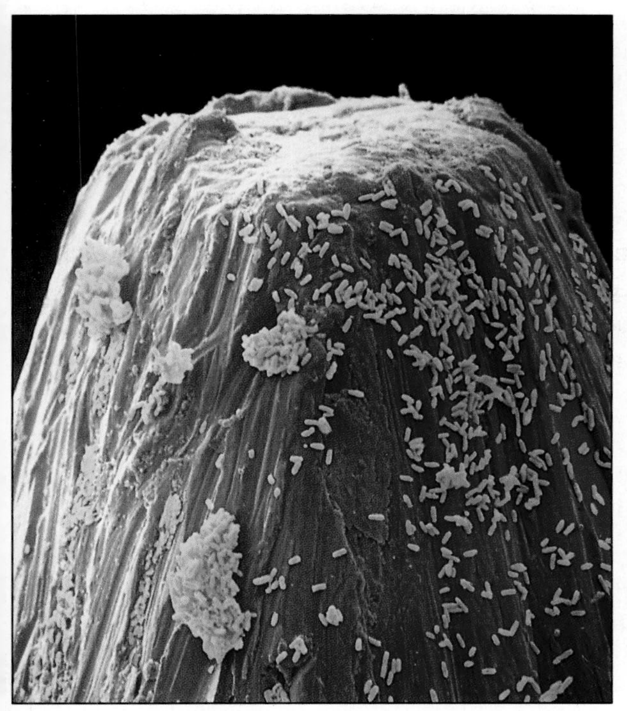

▲ Bacteria on a pinpoint

What are bacteria?

Bacteria are **microbes** – living things so small that you can only see them by using a powerful microscope. They are found everywhere. Most bacteria will not harm us. Some are useful to us such as those we use in yoghurt- and cheese-making. Some bacteria make vitamins to help us digest food. Other bacteria cause food to rot.

Food spoilage

Rotting by bacteria can lead to food spoilage, as food goes 'bad'. **Food spoilage bacteria** do not usually cause food poisoning, although, if there are enough of them, they can make us feel ill.

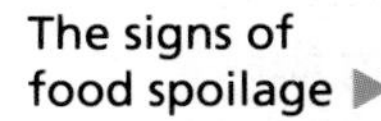

The signs of food spoilage ▶

Food poisoning

Only a very small number of the millions of different bacteria around us are harmful. These are called **pathogenic bacteria**.

◀ You cannot detect pathogenic bacteria by taste or smell. Dangerous food can look perfectly normal.

KEY WORDS

Contaminated – food is contaminated if it contains something which shouldn't be there, eg a stone, chemicals, quantities of harmful bacteria.
Symptoms – the outward signs of an illness, eg sickness, temperature.
Bacteria – single-celled microbes (microscopic living things). Bacteria are found all around us and can survive under many conditions.

Microbe – the name for a microscopic living thing. Bacteria, fungi and viruses are all microbes.
Food spoilage bacteria – bacteria which cause food to decay (rot).
Pathogenic bacteria – bacteria which cause disease.

How bacteria multiply

Bacteria multiply by splitting into two. This is called **binary fission**.

Give bacteria food, warmth, moisture and time, and each one will soon become millions. Some bacteria only take between 10 and 20 minutes to divide.

It may take only nine hours for one bacterium to become 100 million. You still may not see them, as all 100 million would only be the size of a pinhead!

Bacteria need all four to multiply

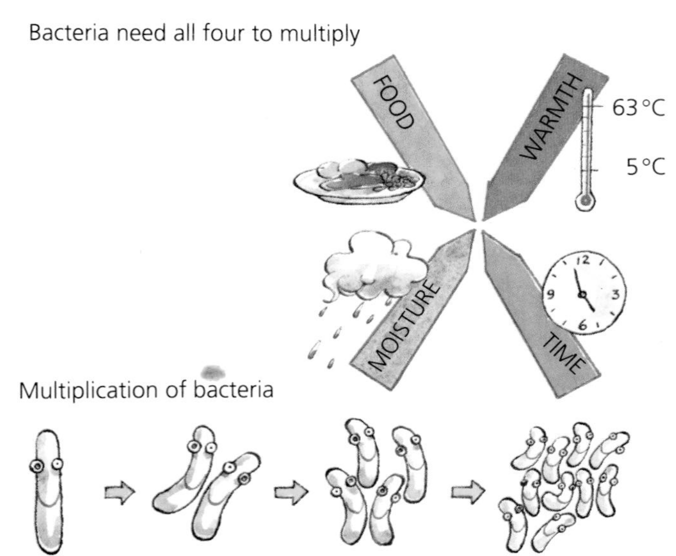

Multiplication of bacteria

Why bacteria make you ill

Some bacteria have to be inside your body to make you ill. Once inside, the bacteria themselves may attack your body causing illness. Some produce a **toxin** (poison) on the food which makes you ill. You can start to feel ill very quickly.

How to detect bacteria

Bacteria are so small that they can only be found with laboratory tests. Scientists take samples of the food, give them warmth, food, moisture and time and let the bacteria multiply until the groups of bacteria can be seen. They can then find out what type of bacteria they are.

The main food poisoning bacteria

There are three common food poisoning bacteria. These are responsible for most cases of food poisoning.

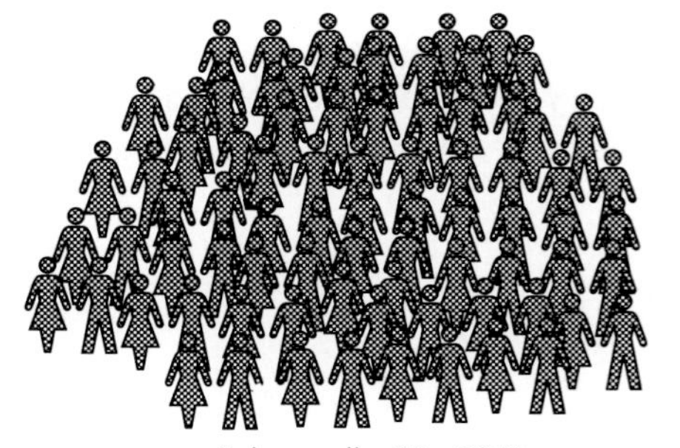

Salmonella 80–90% (mainly from eggs and poultry)

Clostridium perfringens 5–15% (from soil)

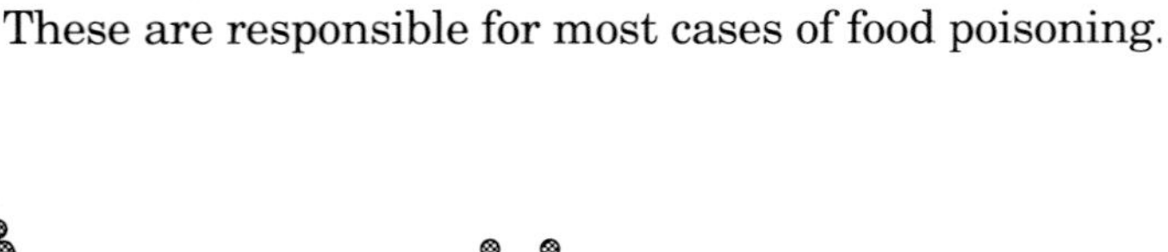

Staphylococcus aureus 1–4% (from coughs and sneezes – ear-nose-throat)

Type of food poisoning	Where the bacteria come from	Onset time	Symptoms
Salmonella	Raw meat, eggs, poultry, animals	6–72 hours	Abdominal pains, diarrhoea, fever, vomiting, dehydration
Clostridium perfringens	Raw meat, soil, excreta, insects	8–22 hours	Abdominal pain, diarrhoea
Staphylococcus aureus	Skin, nose, boils, cuts, raw milk	1–6 hours	Vomiting, abdominal pains, lower than normal temperature

Other bacteria which cause food poisoning include Clostridium botulinum and Bacillus cereus.

Are some foods more risky?

Some foods are **high-risk foods**. This is because bacteria can easily grow on them and they may be eaten without further cooking.

High-risk foods usually are:

- moist
- high protein.

High-risk foods include:

- cooked poultry
- cooked meats
- dairy produce (milk, cream, etc)
- soups, sauces and stocks
- shellfish, seafood
- cooked rice
- raw eggs in food such as mayonnaise or mousse.

Low-risk foods are:

- dried or pickled food
- food with a very high sugar content, eg jam
- food with a very high salt content, eg bacon
- chemically-preserved food.

Bacteria can't multiply easily in these foods.

VIRUSES

Some viruses, which are even smaller living things, can cause food poisoning. They live inside cells and are found directly in food. Again they can only be detected using laboratory tests.

FOOD-BORNE DISEASES

Food-borne diseases are diseases carried by food and water. These diseases are also caused by bacteria and other microbes such as viruses but they are different because they are harmful in small numbers. Campylobacter enteritis is a common food-borne disease. It causes more illness than Salmonella food poisoning. Other diseases include: typhoid, paratyphoid, cholera and dysentery. Contaminated water usually causes these diseases. Brucellosis and tuberculosis are two other diseases which can be caught from drinking untreated milk from infected cows.

KEY WORDS

Binary fission – the process by which a bacterium splits into two.
Toxin – a poison. Some bacteria can produce toxins which make us ill.
High-risk food – food which will support the growth of bacteria easily and won't be cooked further.
Low-risk food – food unlikely to cause food poisoning.

SUMMARY

Food poisoning is caused by eating food which is contaminated.

Most contamination is caused by bacteria. Contamination can also be caused by viruses, chemicals or poisonous plants.

Food poisoning bacteria can make us ill.

The symptoms of food poisoning are abdominal pains, diarrhoea, vomiting, nausea and fever.

Food poisoning can last for days, and can kill.

Food poisoning bacteria become dangerous when they are allowed to multiply.

Food poisoning bacteria can only be detected in a laboratory, they do not usually affect the taste or smell of the food.

Bacteria need warmth, food, moisture and time to multiply. They multiply by binary fission (splitting into two). One bacterium may take only ten minutes to split into two and nine hours to become 100 million bacteria.

The three main food poisoning bacteria are Salmonella, Clostridium perfringens and Staphylococcus aureus.

High-risk foods include cooked meat, cooked poultry, dishes containing eggs, soups and sauces, shellfish and cooked rice.

Food-borne diseases are carried by contaminated food or water. Even a small number of bacteria can make you ill.

Food spoilage bacteria cause food to rot.

So how can you make sure that food does not make people ill? Now read chapter 3 …

CHAPTER 3

How food is contaminated

The best way to stop food getting contaminated is to know how bacteria or unwanted objects get into it. This can happen at any time during food production. Contamination is usually accidental, although it can sometimes be deliberate.

Poison in baby food

POLICE alerted mothers last night after poison and drawing pins were found in a baby food jar.

SLOPPY HYGIENE STANDARDS LEAVE DANGEROUS BACTERIA IN AIRLINE MEALS

Flying fears? It's the food you should worry about

By PAUL CROSBIE
Consumer Editor

AIRLINE food can seriously damage your health, a new investigation reveals today.

Passengers are being exposed to the risk of poisoning because of po hygiene and slo preparation, says Consumer Asso

New storm breaks on food poisoning

Salmonella cases reported in the first three months of this year.

But a separate report circulated to 6,000

Fear over salmonella after death of woman

A woman patient, Laura Smith, has died of salmonella food poisoning at the Ely hospital for the mentally handicapped in Cardiff as concern

FISH FINGERS POISON ALERT

RKETS cleared their stocks of yesterday following a warning they were contaminated with

mous caller said batches of the ourite had been inj with any

SOS OVER TV SWEET SPIKED WITH GLASS

THE LAW

It is an offence to sell or offer for sale

i) food that has been made harmful by the addition or removal of certain substances,

ii) food that is unfit to eat,

iii) food that is so contaminated that it would be unreasonable to expect people to eat it.

(Food Safety Act 1990)

WHERE DO BACTERIA COME FROM?

▲ Bacteria can get on to food from many things during production and preparation.

1 People

Bacteria are often passed from people to food.

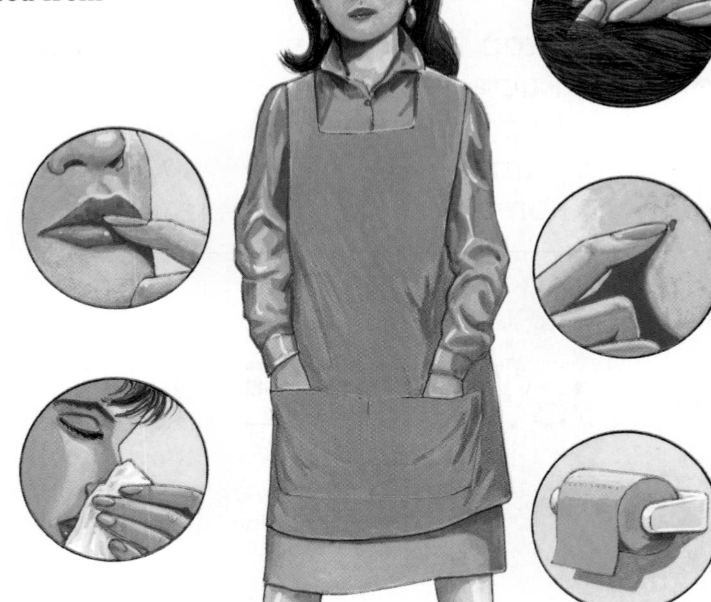

Ears, noses and throats often contain food poisoning bacteria. You can pass them on if you touch your ear, nose or mouth before you handle food.

Coughs and sneezes do spread diseases! (And so does spitting.)

Hands (especially under your nails) can carry bacteria which will be passed on when you touch food.

Hair – if you scratch your head, or let hair droop over the food, dandruff or bacteria may get on the food from your hair.

Spots are caused by bacteria. Scratching your skin will release the bacteria which can then get on to food.

Your intestines contain harmful bacteria which can get on to your hands when going to the toilet. Bacteria can then get on to food from your hands. It is important to remember that toilet paper is porous (bacteria can pass through).

2 Air

Air contains bacteria, and some can settle on food which is left out uncovered.

3 Raw foods

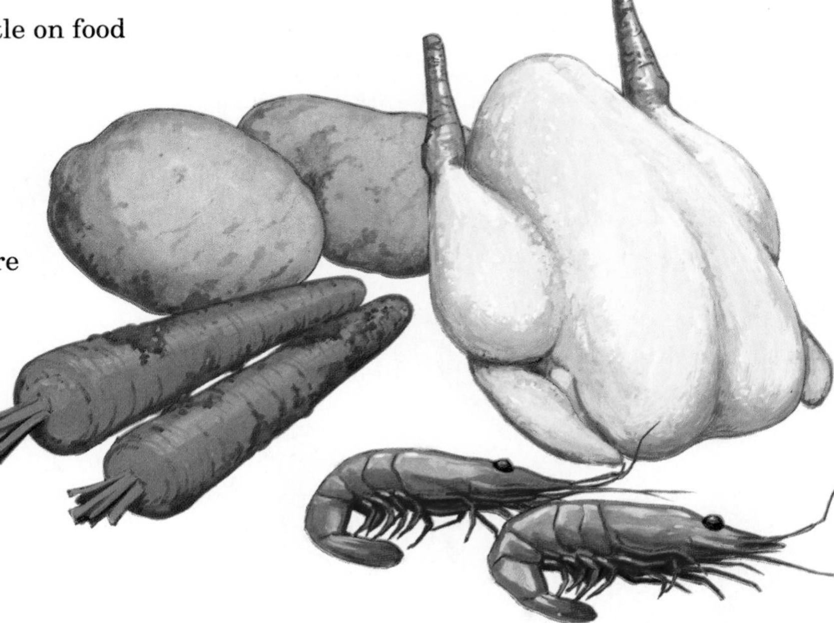

Raw foods have some bacteria on them when they come into food premises.

Meat may have bacteria on it which were in the gut of the animal. Probably the meat was contaminated with bacteria when the animal was killed. Juices from meat may also contain bacteria.

A food poisoning bacterium called Salmonella lives in the gut of many chickens. Most poultry is likely to be contaminated.

Most fruit and vegetables have bacteria on them. Root vegetables such as potatoes, carrots and swedes have soil bacteria on them.

If fish is really fresh, it may not contain many bacteria. However, it goes 'off' very quickly and it is likely to be contaminated during transport and storage.

Shellfish feed by filtering sea water. If shellfish are caught in polluted water, eg near a sewage outlet, they may contain harmful bacteria and viruses.

4 Animals

All animals carry dust, dirt and microbes. This includes household pets such as cats and dogs. If animals are allowed near food, it could become contaminated.

Mice and rats can pass bacteria on to the food in food stores. Rats are very dangerous as they can easily pick up harmful bacteria from the sewers and drains where they live.

5 Insects

Insects carry bacteria on their bodies. Crawling insects such as cockroaches and beetles can get into food stores and contaminate the food.

Flies are very dangerous. They can act as a **vector** of disease. This means that they can carry bacteria from one place to another without being ill themselves. Flies feed on many things (eg manure and rubbish tips) and pick up bacteria from them.

Flies feed by squirting digestive juice on to the food. The food dissolves, then the fly sucks it up. As it feeds, the bacteria from the fly's last meal is squirted over the food. Flies also leave droppings on the food as they eat. Flies do all this very quickly.

▲ Flies carry bacteria.

6 Refuse (rubbish)

Rubbish, especially kitchen rubbish, contains rotting food, leftovers, peelings – all containing bacteria. Open bins will attract flies and other pests which then carry the bacteria to food.

7 Dust and dirt

Dust and dirt is made up from soil, dead skin, fluff and other small particles. It is easily blown on to food after being carried into the kitchen on clothes and shoes. Soil contains the food poisoning bacterium Clostridium perfringens as well as many others.

8 Water

If you use water that hasn't come from a mains tap (eg an uncovered water tank in the loft) it could contain harmful bacteria.

HOW DO BACTERIA GET ON TO FOOD?

Bacteria can't move on their own, so they need a way to get on to food.

Direct contact

Bacteria can be carried from one food to another when the two come into contact.

Indirect contact or cross-contamination

Bacteria can use another object to 'piggyback' on to the food. The object is called a **vehicle**.

If a knife is used to cut up raw meat and then used to cut cooked meat, the bacteria from the raw meat will be picked up on the knife and transferred to the cooked meat. The knife is the vehicle for the bacteria to move from the source to another food.

Examples of vehicles of contamination include:

- containers
- cutlery
- hands
- work surfaces
- cloths
- chopping boards
- equipment.

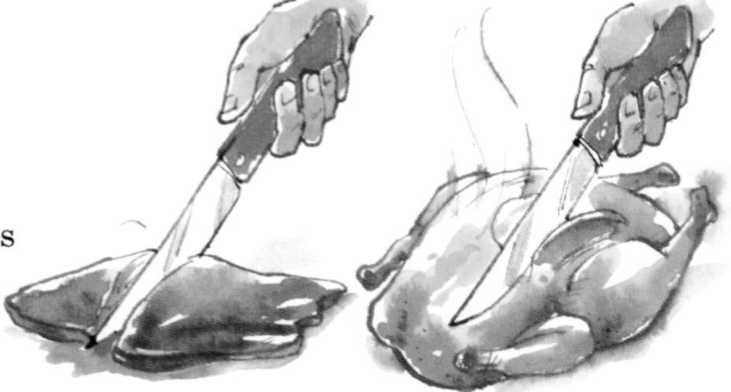

Physical contamination

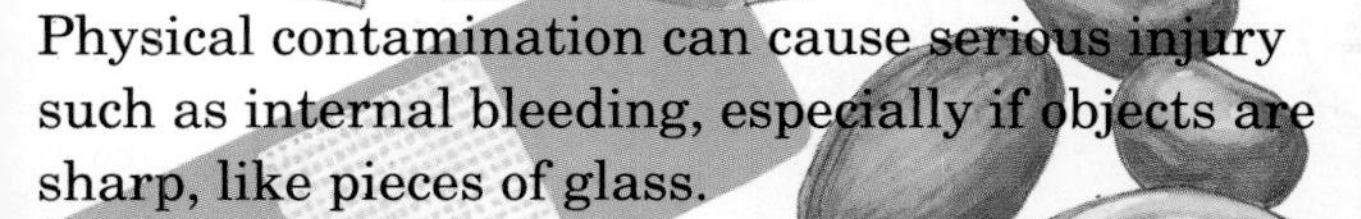

Physical contamination can cause serious injury such as internal bleeding, especially if objects are sharp, like pieces of glass.

Objects get into food during:

- picking – stones and twigs can be found in vegetables
- manufacture – nuts, bolts and other pieces of machinery can drop off machines – production lines often have metal detectors which reject food if there is anything metallic in it
- food preparation – stones, pips, bones and parts of shell in food dishes
- food handling – jewellery, cutlery, money, cigarette ends and plasters have all been found in food.

Deliberate contamination

Sometimes, objects or chemicals are put into foods deliberately to cause harm. This is against the law. Deliberate contamination can ruin businesses as people become worried about the food being dangerous.

THE LAW

It is an offence to sell food that is not of the nature, substance or quality that the customer wants.

(Food Safety Act 1990)

Chemical contamination

Pesticides are sprayed on fruit and vegetables as they are growing and may still be present when harvested.

If fish are caught in polluted water, then they may have harmful metals or other chemicals in them.

Bleach and other cleaning fluids can contaminate food. If you store cleaning fluids in food cupboards, the food may be contaminated, or become tainted.

Some plastic containers release chemicals into food. Only use proper food storage containers.

KEY WORDS

Vector – something which can carry disease-causing bacteria from one place to another.
Cross-contamination – where harmful bacteria are passed from raw food to high-risk food via, for example, a work surface.
Vehicle – an object which bacteria will travel on to get from a source of bacteria on to food, eg hands, knives, cloths.

SUMMARY

Sources of bacteria are people, air, raw foods, animals, insects, refuse, dust and dirt and water which isn't for drinking.

People carry bacteria in their hair, ears, nose, throat, intestines (guts) and skin, particularly hands.

Coughing, sneezing and spitting pass on bacteria. Scratching spots will spread harmful microbes.

Raw foods likely to contain bacteria are meat, poultry, fruit and vegetables (especially soil vegetables), fish, shellfish.

Bacteria from any source can be passed on to food by direct contact.

Work surfaces, knives, cloths and unwashed hands are vehicles for passing bacteria on to food (indirect contact).

Objects can contaminate food during any stage of its production.

Chemicals, including pesticides, bleach and other cleaning materials, can contaminate food if not used carefully.

If harmful objects are put in food on purpose, this is called deliberate contamination and it is a criminal act.

How do you prevent food from being contaminated? Now read chapter 4 ...

4

CHAPTER

Preventing food poisoning

Bacteria cause most food poisoning. The three ways to stop food poisoning are to stop bacteria getting on to the food, to stop the bacteria on the food from multiplying to a dangerous level and to destroy the bacteria on the food.

PROTECTING FOOD AGAINST CONTAMINATION

▲ Covering foods

1 Get your food from a safe source, so that it isn't likely to be contaminated in the first place.

2 Avoid contamination by:

▼ Handling food as little as possible

▲ Making sure raw and cooked foods are kept separate

▼ Keeping animals out of food workplaces

▼ Getting rid of rubbish carefully in covered bins

◀ Keeping work-places and workers clean

▶ Getting into good habits and keeping yourself clean

3 Avoid opening up routes for bacteria so:

- only use knives and spoons once, then wash them
- regularly wipe and disinfect work surfaces, chopping boards and only use disposable cloths
- use separate equipment for raw and cooked food.

Keeping food away from sources of bacteria is not enough on its own. Many raw foods already contain bacteria. You must make sure the bacteria do not multiply.

THE LAW

Persons handling food must protect it from contamination and not place it where there is a risk of contamination.

(Food Hygiene (General) Regulations 1970 as amended)

PREVENTING MULTIPLICATION OF BACTERIA

Bacteria need four things to be able to multiply. These are food, warmth, moisture and time. By taking one of those four away, you can slow down or even stop bacteria from growing. Once bacteria are on the food, how can we stop them multiplying?

Warmth

The effect of temperature on bacteria:

Cooling bacteria slows down the multiplication. Heating to above 70 °C for sufficient time will kill most bacteria.

Beware the **danger zone** of 5–63 °C!

Foods are at this temperature:

- when they are left out in a warm room
- when they are being slowly heated up
- when they are cooling down after cooking
- if left in the sun in shop windows
- if hot sauce/gravy has been poured onto cold food.

Avoiding danger

Do not keep foods at room temperature longer than necessary. Either keep food hot or cold.

Store foods at temperatures low enough to prevent the bacteria from multiplying.

Cool cooked food completely before refrigerating and for not longer than 1½ hours. Cover food as it is cooling.

Keep hot foods very hot before serving, and try not to prepare foods too long in advance.

Bacterial spores

Some food poisoning bacteria, eg Clostridium perfringens form spores to protect them if they are in conditions that they don't like. Spores are hard, tough cases and they can survive high temperatures, drying and disinfectants. When the conditions improve, the bacteria are released from the spores and can then multiply.

THE LAW

By law, certain foods must be kept cold (less than 8 °C) or kept hot (at least 63 °C). From April 1993, some of these foods must be kept extra cold (below 5 °C) eg soft cheeses, ready-to-eat cooked products, smoked meat or fish.

(Food Hygiene (Amendment) Regulations 1990 and 1991)

KEY WORDS

Danger zone – the temperature range between 5 °C and 63 °C in which bacteria multiply easily.
Spores – hard cases which protect certain bacteria if they are in conditions that they don't like. Resistant to heat, cold, drying and disinfectants. Bacteria come out of the spores when conditions get better.

	Temperature	Conditions	Bacterial action	Safety
	–18 °C	Freezers	Dormant – not able to multiply	Safe
	1–4 °C	Fridges and cold stores	Most bacteria unable to multiply	Safe
	5–63 °C	Room temperature (10–36 °C) Body temperature (37 °C) Warm food (38–63 °C)	Bacteria able to multiply	DANGER
	64–72 °C	Keeping food hot	Most bacteria can't multiply	Safe
	73–100 °C	Cooking temperature	Most bacteria die	Safe
	Above 100 °C	Boiling food Pressure cookers	Most bacteria and bacterial spores killed	Safe

Moisture

Bacteria need moisture (water) to multiply.

Food that does not contain water, such as rice, pasta, flour and dried fruit does not allow bacteria to multiply.

Once water is added to dried food, for example baby food, bacteria could multiply.

Time

The longer food is left, especially at dangerous temperatures, the more time the bacteria will have to multiply.

Use fresh food as soon as possible.

Prepare food just before it is needed.

If all the food isn't used, refrigerate it as soon as possible. If you reheat food later, you must heat it thoroughly to a high temperature. Ideally, food should not be reheated.

▲ Preserved foods.

If frozen food is allowed to thaw, the bacteria in the food will start to multiply. Once frozen food has thawed, you mustn't refreeze it.

Food can be **preserved** by making the conditions unsuitable for bacteria. This keeps the food safe for longer. Salting, pickling and jam-making all work by stopping the bacteria multiplying.

DESTROYING THE BACTERIA IN THE FOOD

The best way to destroy bacteria is by heating. This process is also used to preserve some foods.

Sterilisation

Food is heated to a high temperature for a long time. This kills all the bacteria and then the food is sealed in airtight containers. Milk is sterilised at 150 °C for 2 seconds. The high temperature does cause changes in taste and texture.

Pasteurisation

Food is heated up to 72 °C for 15 seconds, which kills pathogenic bacteria, but not all spoilage bacteria. Bacterial spores will survive, so pasteurised food must be kept in refrigerators.

Ultra Heat Treatment (UHT)

The food is heated to a very high temperature for a short time and nearly all bacteria are killed. This changes the taste and texture less than sterilisation.

Cooking/processing

Always thaw frozen food properly before cooking unless the instructions tell you to cook it from frozen. If you don't, the food may be cooked on the outside, but the temperature in the middle may still be in the danger zone.

Cook food for a long enough time. Larger items of food, eg joints of meat, need longer cooking. This will make sure that it is cooked all the way through. You should not use joints larger than 2½ kgs (6 lbs). If a joint is larger then cut it in two.

▼ When you cook food, you must make sure that the food gets hot enough and is cooked for long enough.

THE TEN MAIN REASONS FOR FOOD POISONING

1	Food is prepared too far in advance, and stored at a warm (dangerous) temperature.
2	Food is cooled too slowly before being refrigerated.
3	Food isn't reheated enough to kill all the bacteria in it.
4	People eat cooked food which has been contaminated by food poisoning bacteria.
5	Food is undercooked.
6	Poultry is not thawed properly.
7	Cooked food is cross-contaminated by raw food.
8	Hot food is kept warm at a temperature of less than 63 °C.
9	Food handlers pass on infections when handling the food.
10	Left-overs are used.

Source: Dr D Roberts J. Hyg, Cambs. (1982) VOL 89, pp491–498

KEY WORDS

Preserve – to treat foods so that they last longer than they would normally.

SUMMARY

Prevent food from becoming contaminated

- Handle food as little as possible
- Keep food away from all sources of bacteria
- Cover food
- Keep raw and cooked foods separate
- Keep all animals and insects away from food places
- Dispose of waste food and other rubbish carefully
- Keep bins covered
- Keep everything as clean as possible.

Stop the bacteria on the food from multiplying

- Prevent dry foods from becoming moist. Bacteria cannot grow without moisture
- Store food at safe temperatures – keep cold food below 5 °C. Keep hot food above 63 °C
- Cook food thoroughly
- Try not to prepare food in advance
- Do not keep food in the temperature danger zone (5–63 °C) for any longer than necessary
- Avoid reheating food.

Remember, bacteria need food, moisture, warmth and time to grow.

Destroy the bacteria in the food

- Heat food thoroughly
- Thaw frozen food thoroughly before cooking unless the instructions say otherwise.

Keeping your workplace clean prevents food contamination.
Chapter 5 tells you how to keep clean ...

Cleaning the workplace

Food places must be kept clean and tidy, and they must be disinfected regularly. It is important to know how to clean everything properly so that it is safe.

EFFECTIVE CLEANING

Cleaning is hard work and requires energy. For cleaning to be most effective, you should use hot water, a **detergent**, and some physical effort. A detergent is a chemical which helps to dissolve grease and remove dirt.

Even if a surface looks clean, it may still have bacteria on it. To make sure that it is safe, it also has to be **disinfected**.

Disinfection is the reduction of bacteria to a safe level. The most common ways of disinfection are by using hot water (from 82 °C), steam or a suitable disinfectant.

▲ Things you need for effective cleaning.

Before you start cleaning, put away or cover all food.

To clean things properly, you should follow six stages:

The pre-clean removes the worst of the dirt or left-over food before the main clean. The main clean should be carried out using hot water and detergent.

It is important to remove all traces of detergent before using disinfectant, so there must be a rinse. As the item will now be clean, it can be disinfected, then any traces of the disinfectant can be removed by a final rinse in clean, hot water. The item should then be left to dry.

▲ The six stages of cleaning and disinfection.

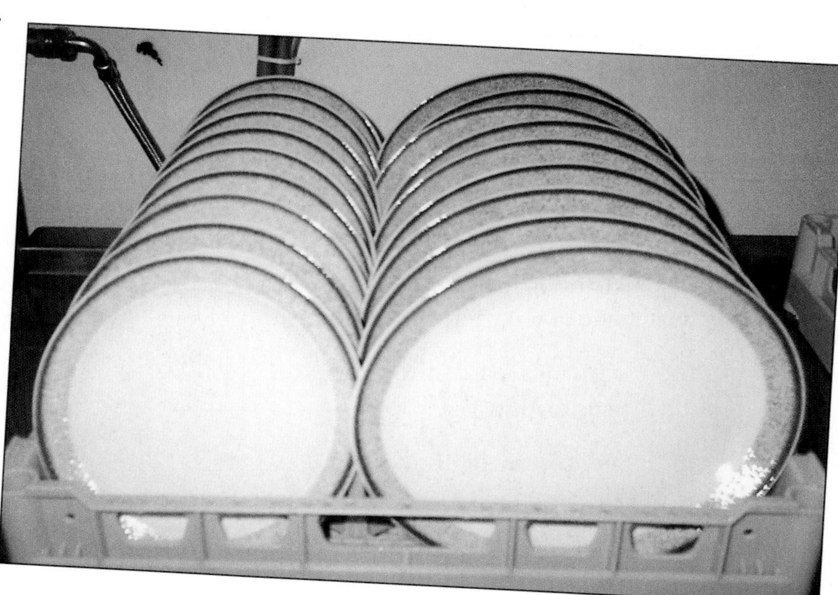

▶ It is best to let objects dry on their own, as drying cloths can spread bacteria. If cloths have to be used, they should be made of disposable material (paper) or should be clean and dry.

A **sanitiser** is able to clean and disinfect. It is a chemical which has both detergent and disinfectant in one solution, so it combines stages 2, 3 and 4. Do not try to combine these two chemicals into one yourself.

The six stages of cleaning can be applied to anything in a food workplace. Here is an example:

Work surfaces

1	Pre-clean	Brush away food debris and wipe surfaces. Use scourer to loosen stubborn spots.
2	Main clean	Use clean water and detergent. Pay attention to difficult areas, eg corners.
3	Rinse	Use clean water and a clean cloth.
4	Disinfection	Use a disinfectant solution and leave on for contact time.
5	Final rinse	Use clean water and a clean cloth.
6	Drying	Leave to dry naturally.

If a sanitiser is used, stages 2, 3 and 4 are combined.

KEY WORDS

Detergent – a chemical which dissolves grease. An aid to cleaning.
Disinfection – reducing the number of bacteria to a safe level.
Disinfectant – a substance which is used for disinfection.
Sanitiser – a chemical which cleans and disinfects.

Washing-up

Washing-up using a double sink:

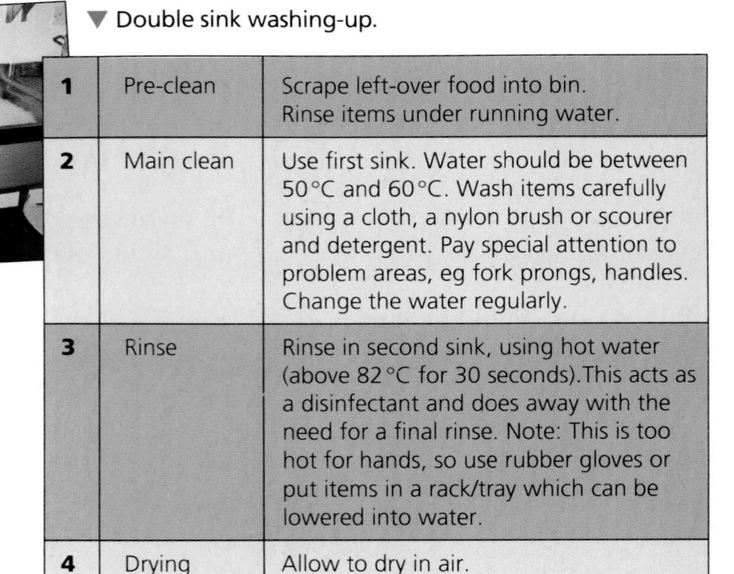

Effective washing up needs hot water, detergent, disinfectant, drying facilities. ▶

▼ Double sink washing-up.

1	Pre-clean	Scrape left-over food into bin. Rinse items under running water.
2	Main clean	Use first sink. Water should be between 50 °C and 60 °C. Wash items carefully using a cloth, a nylon brush or scourer and detergent. Pay special attention to problem areas, eg fork prongs, handles. Change the water regularly.
3	Rinse	Rinse in second sink, using hot water (above 82 °C for 30 seconds).This acts as a disinfectant and does away with the need for a final rinse. Note: This is too hot for hands, so use rubber gloves or put items in a rack/tray which can be lowered into water.
4	Drying	Allow to dry in air.

Washing-up using a dishwasher can be more efficient as the water temperature is easier to control. Check the temperature is hot enough and that there is enough detergent.

THE LAW

Sinks for washing food or equipment must be provided and kept clean. Hot water at a suitable temperature or hot and cold water must be supplied.

(Food Hygiene (General) Regulations 1970 as amended)

CLEAN AS YOU GO

All staff should know about the cleaning schedule. This tells you:

- how to clean
- what chemicals to use
- how often to clean
- who should clean.

Where possible, it is important to clean as you go. All workplaces should be left tidy, clean and disinfected at the end of each working day. You should disinfect the following often:

- food contact surfaces, eg chopping boards
- hand contact surfaces, eg fridge door handles.

Clean from the cleanest area to the dirtiest area to avoid contamination.

Chemicals such as detergents and disinfectants should be used carefully. Always follow the manufacturer's instructions about how much to use. This is important as:

- too little may mean that the chemical won't be effective at the job
- too much does not mean that it is more effective – it could be difficult to rinse off and will be wasteful.

Disinfectants must be left on surfaces long enough to work properly. Remember to follow the manufacturer's instructions. Always use fresh disinfectant each time you clean. Do not soak mops or cloths overnight. After cleaning and disinfecting, the cleaning equipment should be allowed to dry and should be stored away from food.

▲ Don't be like this – clean as you go.

KEY WORDS

Clean as you go – the routine of cleaning up as you work, not leaving all the cleaning to do at the end.

FOOD-HANDLING EQUIPMENT

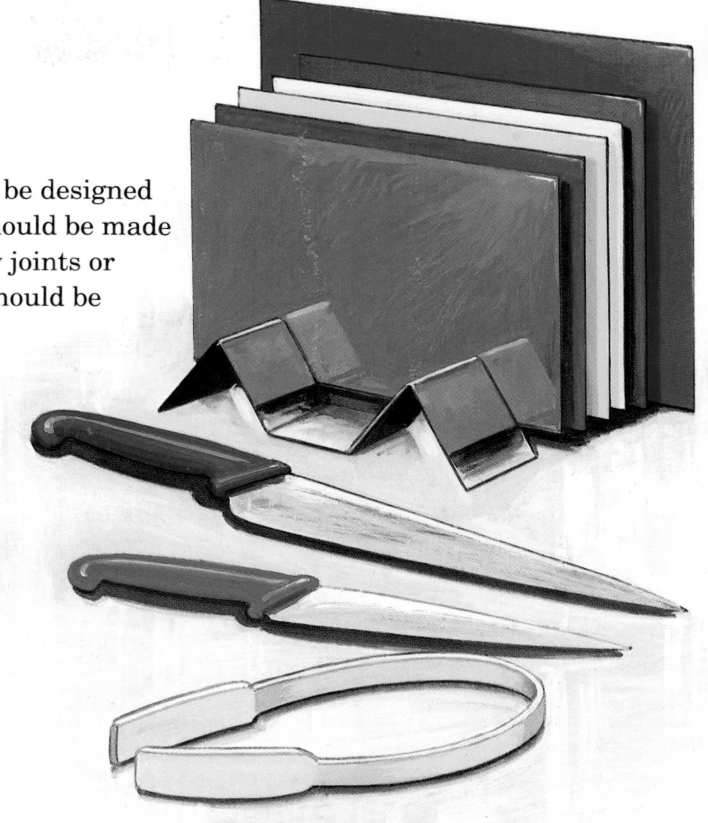

Equipment which you use with food should be designed for easy cleaning. It should be smooth. It should be made of non-reactive material. It should have few joints or holes where dirt and bacteria can hide. It should be easily movable.

Equipment should be made from materials which are:

- resistant to rusting
- non-toxic
- non-absorbent
- durable (long lasting), eg stainless steel, polypropylene. Don't use wood – it absorbs liquid and is difficult to clean.

There should be enough equipment available in food preparation areas so that no one has to share. This lessens the risk of food contamination.
All equipment should be made of non-toxic materials which are hard-wearing.

Cutting boards should be made of polypropylene or another hard material which is easy to clean. They should be in one piece, with no joins or cracks. It is best if they are colour-coded, so that boards for raw foods can be kept separate from those used for cooked foods. Cutting boards should be washed and disinfected straight after they have been used.

Knives should be made from high-quality food-grade stainless steel, with moulded polypropylene or hard plastic handles. The handles could be colour-coded in the same colours as chopping boards to avoid cross-contamination.

Try to use tongs when you handle food. One pair of tongs should be used for one kind of food.

Cloths should be either disposable or they should be used only once before being washed and disinfected.

SUMMARY

Cleaning needs physical effort, hot water and a detergent to remove dirt.

After anything has been cleaned, it must be disinfected. This can be done with hot water (above 82 °C), steam or a disinfectant.

There are six stages to cleaning:

1. Pre-clean to remove the worst of the dirt
2. Main clean
3. Rinse to remove detergent from main clean
4. Disinfect to reduce the number of bacteria to a safe level
5. Rinse
6. Dry.

Where possible, leave things to dry on their own.

Follow the manufacturer's instructions when using chemicals.

Put all food away before cleaning the workplace.

All equipment used during food preparation should be made of a suitable material which is easy to clean and disinfect.

Clean as you go. Always use fresh disinfectants and leave them on surfaces for long enough to take effect.

Equipment should be colour-coded to avoid cross-contamination.

Tongs should be available to avoid unnecessary food handling.

Your workplace may be clean, but what about you? Read chapter 6 ...

6

CHAPTER

Personal hygiene

If you are working with food, it is essential that you keep yourself clean and wear clean protective clothing. It is important to have high standards of personal hygiene when working with food.

ARRIVING FOR WORK

Leave your outdoor clothes away from food preparation areas. Change your shoes so that soil and dust can be kept out of the work area.

THE LAW

Accommodation for outdoor clothing must be provided.

(Food Hygiene (General) Regulations 1970 as amended)

HAIR

Hair contains bacteria. It falls out naturally – you can lose up to 100 hairs a day. This makes your head a likely source of contamination.

Hair should be clean, tied back and covered when you are handling food. Never scratch your head when there is food around. Comb your hair before you put your workclothes on. Never comb your hair in a food preparation area.

MOUTH, NOSE AND EARS

Many people have Staphylococcus aureus living in their mouths, noses and ears. To avoid transferring these harmful bacteria onto food in a food room:

▲ DON'T cough or sneeze over food

▲ DON'T taste food with your fingers

▲ DON'T bite your nails or lick your fingers

▲ DON'T pick your nose

▲ DON'T spit

▲ DON'T eat or chew gum

▲ DON'T smoke

Smoking is illegal in food rooms. It can lead to ash or cigarette ends in the food and it could make the food smell. Smoking also causes coughing which spreads bacteria. When you smoke you pass harmful bacteria from your mouth on to your fingers which can then be spread on to food.

If you need to blow your nose, do it away from food; use a paper tissue then throw it away immediately. Wash your hands straight away.

THE LAW

It is illegal to smoke in food rooms.

(Food Hygiene (General) Regulations 1970 as amended)

SKIN

Even clean skin has a lot of bacteria living on it. Spots are caused by bacteria in the pores of your skin. Body odour is caused by bacteria living on stale sweat. You must wash regularly to remove these bacteria.

Never scratch your skin, especially spots, as this will leave bacteria on your hands which can then be passed on to food.

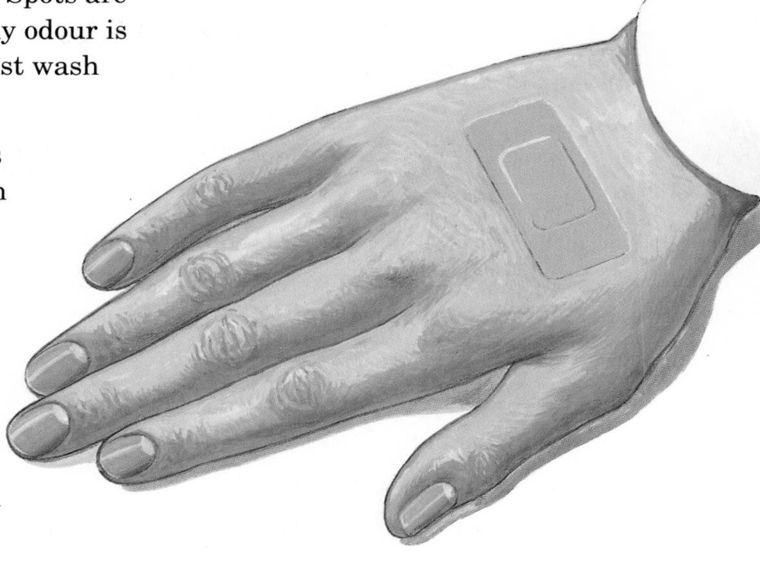

Cuts and grazes

Any skin infection is likely to contain Staphylococcus aureus, one of the three most common food poisoning bacteria (see page 10).

All cuts and wounds should be covered with a waterproof dressing before starting to work with food. This protects the food from contamination. The dressing should be brightly coloured– blue is usual – so that it can be seen easily if it falls off.

Some dressings have a metal strip so they can be detected by a metal detector on production lines.

If a dressing does fall into food, the food must be thrown away.

THE LAW

Food handlers must keep clean and cover any cuts with waterproof dressings.

(Food Hygiene (General) Regulations 1970 as amended)

HANDS

Wash your hands often, and always after

- visiting the toilet
- touching your face – especially your nose, mouth and ears
- handling raw food
- handling rubbish
- breaks, and any time you come back into a food preparation area
- cleaning and disinfecting the workplace.

Hands are the parts of the body which come into contact with the food most often. It is essential that they are properly washed before work and frequently during work.

Your nails should be kept short. Nail varnish can chip off and can also hide dirty nails so it shouldn't be worn for work.

By law, all food preparation areas must have separate basins for washing hands. There must be a good supply of soap, hot and cold (or warm) water, a clean nailbrush and a hygienic way of drying your hands.

Don't wear rings or watches for work – bacteria can live on watch straps and under rings too.

It is best to use liquid soap – bars of soap may have bacteria on them from the last person.

Wash your hands and wrists using hot water, either in a basin of clean water, or under a running tap. Use a nail brush to clean under your nails and between your fingers.

Dry your hands carefully using a hot air drier or a clean towel. Paper towels are best as they are used only once and won't pass on bacteria from other people.

Washing your hands often during food handling keeps the number of bacteria down and prevents cross-contamination.

JEWELLERY, MAKE-UP AND PERFUME

Jewellery, including watches, shouldn't be worn whilst working with food. It harbours bacteria and may fall off into the food.

Heavy make-up or strong aftershave or perfume can make food smell.

CLOTHING

Protective clothing must be provided when working with food. You wear protective clothing to protect the food, not your clothes! Make sure that it is clean and preferably light coloured so any dirt will show.

Other things to remember are:

- hats or other forms of head covering should be provided
- clothes should be well fitting for comfort and have no pockets
- only fastenings which will not come off in food, eg velcro, press studs and tapes, should be used
- work clothes must give good protection and cover all other clothes
- always wear clean shoes. Do not wear sandals or open footwear in a kitchen.

▲ Ideal protective clothing.

Other protective clothing includes:

- hair nets
- disposable plastic gloves
- washable head and neck covering if your job involves carrying raw meat carcasses, to avoid contaminating the meat.

THE LAW

Food handlers must wear suitable protective clothing.

(Food Hygiene (General) Regulations 1970 as amended)

ILLNESS

If you have any of the symptoms of food poisoning, or have been in contact with someone with food poisoning, or have an ear, nose or throat infection, tell your supervisor. Your supervisor has to tell the local authority if there is any employee suffering from food poisoning or a food-borne disease. Food handlers who have had food poisoning should not return to work until they have medical clearance.

You must tell your supervisor immediately if you are suffering from:

- a cold
- a sore throat
- boils or spots
- a septic wound
- diarrhoea
- an upset stomach
- sickness.

Some people who have had food poisoning may still have bacteria in their gut even though they no longer have any symptoms. These people are called **carriers**. They can still pass bacteria on to others through poor hygiene during food preparation or processing.

THE LAW

Food handlers must report to their supervisor if they are, or suspect they are, suffering from food poisoning, or a food-borne disease.

(Food Hygiene (General) Regulations 1970 as amended)

KEY WORDS

Carrier – a person who has been ill with food poisoning and may still be carrying food poisoning in their gut.

SUMMARY

Leave outdoor clothes away from food areas.

Cover hair, and don't touch or comb hair when in a food area.

Don't touch skin, especially your mouth, nose and ears.

Cover all cuts and grazes with a clean waterproof dressing.

Don't wear jewellery, watches , heavy make-up or strong perfume.

Wash your hands often.

Wear protective clothing. This should be:

- clean
- washable
- well fitting
- light coloured.

Report any illness to your supervisor.

Never smoke in food rooms.

Now you know how to keep yourself clean, but how should food be stored safely? Read chapter 7 to find out ...

7

CHAPTER

Food storage

You must store all kinds of food safely and at the right temperature to avoid contamination and waste. Storage conditions and the length of time that food can be safely stored depend on what sort of food it is and how it has been preserved.

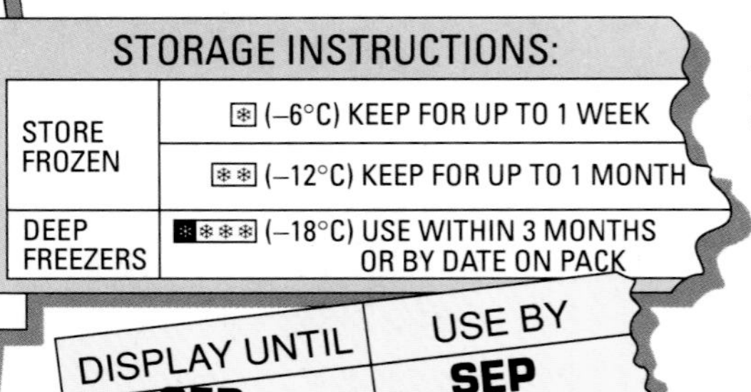

Food is often preserved before being packed. This means that the food will keep longer on the shelf than it would do if it was fresh.

◀ Look for the label that tells you how long you can keep food safely.

For many years, food has been preserved by canning, bottling, smoking and drying.

Modern methods include:

- chemical preservation
- freezing
- vacuum packing
- heat treating (pasteurisation, UHT)
- irradiation (for some foods).

It is important to **rotate stock**. This means making sure that the old supplies of food are used before the new stock is used. Have a routine, eg put the new stock at the back of the shelf and pull the older stock to the front. Make sure that everybody knows the routine and that it is followed. This also helps to avoid running out or overordering.

Dry food storage

Dry foods like flour, rice, pasta and powdered foodstuffs should be kept in rooms or cupboards which are dry, cool, well lit and well ventilated. The food should also be protected against food pests and it should never be stored on the floor. Shelving such as racks made from tubular stainless steel should be used. Put loose powders in plastic bins with airtight lids. If the bins are large, they need to be easy to move.

The place should be kept clean and tidy.

Keep bottled and canned food in the same way. You should throw away any cans which are:

- dented
- rusty
- damaged in *any* way
- 'blown', that is, bulging at the top or bottom
- past their 'Best before' date.

Never open a blown can as it contains a lot of gas which could have been produced by harmful microbes.

▲ Do not use dented or blown cans.

THE LAW

It is illegal for anyone, other than the person originally responsible for packing, to change a date mark after it has been set.

It is illegal to sell any food after its 'Use by' date.

(Food Safety Act 1990)

Vegetables and fruit

Fresh vegetables and fruit should be used as soon as possible. Keep them in a cool room and store them off the floor.

Root vegetables such as potatoes and carrots should be stored away from other vegetables and fruit to avoid contamination of other foods from soil. Root vegetables should be kept in the dark.

▲ Green potatoes can be poisonous.

All fruit and vegetables should be checked regularly as they rot quickly. If any show signs of rotting, they should be removed and thrown away immediately.

Perishable foods

These are foods like meat, fish and dairy products which spoil quickly or may be contaminated with bacteria. If allowed to multiply, these bacteria would cause food poisoning or food spoilage.

Perishable foods should be refrigerated as most bacteria are **dormant** (not able to multiply to dangerous levels) at temperatures of less than 5 °C.

°C: 110, 100, 90, 80, 70, 60, 50, 40, 30, 20, 10, 0, –10, –20, –30

Room temperature	(10 – 36 °C)
Refrigerator	(1 – 4 °C)
Freezer	(–18 °C)

KEY WORDS

Stock rotation – a system where the oldest food is used first. New stock should be placed at the back of, or under, the existing stock.

Dormant – when any bacteria on the food are not multiplying, ie a 'sleeping' stage in which no growth occurs. The bacteria are still alive if they are dormant.

FRIDGES

Site in an area which is well ventilated and away from direct sunlight.

Surfaces must be easy to clean.

Should be made of materials which will not rust or absorb anything.

Service regularly.

Fridge thermometer to be checked every day. Temperature must be between 1–4°C.

Must be regularly defrosted and cleaned – self-defrosting fridges should be cleaned at least once a week. (Use a non-perfumed cleaner so that the food doesn't pick up the smell.)

Air must be able to circulate freely inside so don't overload a fridge.

Door seals should be tight fitting and clean.

Never store food in front of the cooling units – it will stop the air circulating.

Don't leave the door open longer than you need to – keep the fridge cold!

Storing food in fridges

Unpack and store deliveries of refrigerated food as soon as possible.

Wrap food well and place in separate containers to avoid cross-contamination.

Keep cooked foods away from raw foods. Keep raw meat on a bottom shelf and in a container. This stops the juices from running on to other foods. It is best to keep raw meats in a separate unit if possible.

Dairy products should also be kept away from other foods – a separate fridge is best.

Opened cans of food shouldn't be stored as the food can react with the metal of the cans. Empty the food into a suitable container, and cover.

Check perishable food in fridges every day and rotate the stock – first in, first out.

Don't put hot food in a fridge; it increases the temperature of the fridge and causes **condensation**.

Drips of condensation can cause contamination.

KEY WORDS

Condensation – the formation of water droplets from steam, often caused in fridges by hot food being placed inside.
Perishable foods – foods which spoil quickly.

FROZEN FOODS

Delivery

If the frozen food is going to be stored, then check to make sure it is still at a safe temperature of about –18 °C as soon as it is delivered. If it is, put it into frozen storage as quickly as possible. If the food is warmer than –15 °C, don't accept the delivery.

If frozen food is delivered and is going to be used straight away, store in a refrigerator and use it as soon as it is thawed. *Never* refreeze frozen food once it has thawed.

Freezers

The safe temperature for freezers is between –18 and –23 °C.

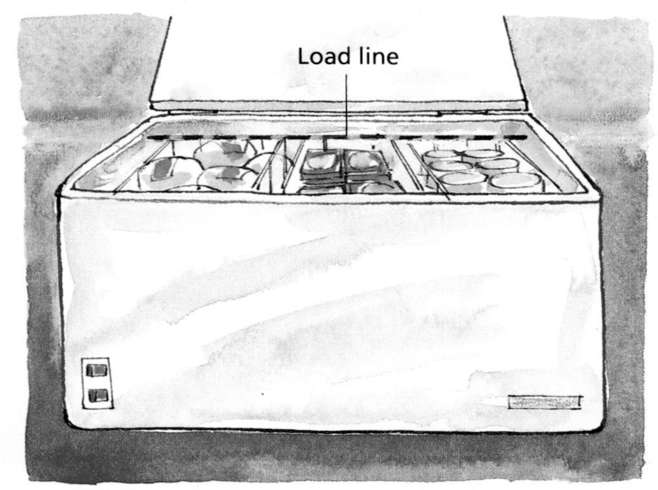

For freezers and frozen food display cabinets, it is important that you don't overfill them above the marked line as the food which is on the top may start to thaw.

Frozen food stays in a good condition because microbes are dormant at very low temperatures. In addition moisture is unavailable to them when it is frozen. This means that they can't multiply.

Rotate stocks of frozen food carefully in freezers as it is easy to leave old stock at the bottom of chest freezers without realising it. Check any date marks before using the food.

Freezer sections in fridges have a star rating. This tells you how long food can be safely kept in that part of the fridge.

❄	one week
❄❄	one month
❄❄❄	three months
❄❄❄❄	three months or longer. Capable of freezing fresh foods

Freezers should be defrosted and cleaned regularly. Do this when stocks are low just before a delivery. Food should be moved to another freezer to keep it frozen whilst you are cleaning.

All freezers, display cabinets and cold stores should be serviced regularly by qualified engineers.

Some frozen foods can be cooked immediately. Look at the cooking instructions carefully before cooking. Other frozen foods such as poultry and meat joints must be thawed completely before cooking. If you don't do this, the food will not cook properly and bacteria could multiply.

SUMMARY

When storing food, it is important to rotate stock so that the oldest food is used first.

Never overstock on any food item.

Follow the manufacturer's instructions when storing all foods. Look for 'Use by' and 'Best before' dates.

Dry, bottled and canned foods should be stored in a dry, well-ventilated room, off the floor and in airtight containers where necessary.

Fruit and vegetables should be stored in a cool dry room or cupboard and checked often. Root vegetables should be stored separately and in a dark place.

Perishable foods should be refrigerated: raw foods, cooked foods and dairy products should be stored away from each other. All foods should be kept in covered sealed containers to avoid contamination.

Refrigerators should be kept at a temperature of 1 to 4 °C.

Freezers should be kept at a temperature of –18 to –23 °C.

So your food is stored safely, but how should you prepare it? Now read chapter 8 ...

Safe food preparation

Preparing food safely means thawing it properly and keeping food out of the danger zone (5 °C – 63 °C). Safe food preparation helps to stop cross-contamination by keeping all equipment clean and storing high-risk foods away from other foods.

Planning

Careful planning of food preparation can save time, energy and food. Make sure that the food to be prepared is correctly thawed, and that you have enough time, space and equipment to do your job properly. Work out how much food will be needed, to avoid unnecessary waste.

Timing

It is important to plan time carefully so that there is the shortest possible time between:

- taking food out of refrigerated storage and preparing, cooking or eating it
- cooking food and eating it
- cooking, cooling and refrigeration.

Defrosting

Some foods can be cooked from frozen, but some have to be thawed completely. Poultry and large meat joints are examples of foods which must be thoroughly defrosted before they are cooked.

If food is still frozen, the outside may cook while the inside may only reach room temperature. This means that any bacteria in the food will be able to multiply to a dangerous level.

To thaw frozen food, put it in a cool area well away from any other food preparation. The area must not be used for cooling cooked items at the same time, or you could get cross-contamination from the raw food to the cooked food.

Thawing poultry

Remove from the freezer, and place in a cool area or thawing cabinet away from all other foods. Put the bird in a dish to catch the juices and then cover it.

Small birds can be thawed at the bottom of a fridge or in a cold store. It will take longer but it is less likely that bacteria will grow.

Remove the giblets – if you leave them in, the bird will take longer to thaw.

Check the bird regularly, and throw away any liquid collected in the dish.

Poultry is properly defrosted when the legs and body are soft and the legs can be moved. There should be no sign of ice inside the body.

Once it has thawed, store in a refrigerator, away from other foods. Cook within 24 hours.

Do not refreeze it.

Clean and disinfect the equipment.

Wash your hands after handling the bird.

▼ Food should be thawed in a suitable container.

◀ This food was condemned by an Environmental Health Officer.

Thawing times for frozen poultry

Weight	Approx. thawing time in a cool room
2.25 kg (5 lb)	15 hours
4.5 kg (10 lb)	24 hours
6.75 kg (15 lb)	36 hours
9.0 kg (20 lb)	48 hours

It is recommended that thawing of larger frozen poultry is carried out at 10–15 °C, entirely separate from other foods or in a thawing cabinet or cool room.

Very large turkeys and geese can take up to three days to thaw completely.

Thawing times for meats

CUT OF MEAT	DEFROSTING TIME
	In fridge (per 0.5 kg (1lb))
Chops, steaks	5–6 hours
Small joints of meat (less than 1.5 kg (3lb)	3–4 hours
Large joints of meat	6–7 hours

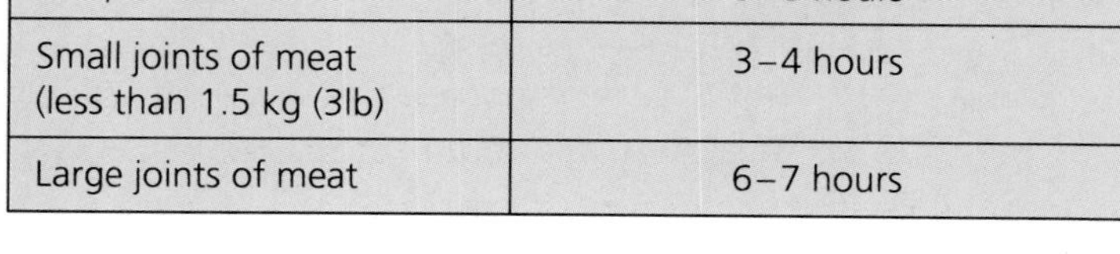

Food preparation

Wash your hands regularly during food preparation. Use the hand washing basin provided, not the sink. Always wash your hands after:

- touching raw foods
- handling waste
- cleaning.

Prepare all raw food away from other foods. Throw away peelings and waste, and clean and disinfect all equipment before you prepare other foods. Clean as you go.

Preparing high-risk foods

Food which will not be cooked or processed any more before it is eaten must be kept away from raw foods. It should be prepared in a separate area, using separate equipment.

Preparing meat

You must use separate equipment when you are preparing cooked meats. Do not use the same knives or slicers on raw meat (including bacon) and cooked meat. If you are preparing canned meat, the can opener, work surface and knife or slicer must be cleaned and disinfected immediately before and after you use them.

Cooking foods

Make sure all foods are cooked through before they are served. It is best to use food thermometers or probes to make sure that foods have reached a safe temperature in the middle. Make sure that thermometers are clean and disinfected.
The maximum joint size should be 2½ kg (6 lb).

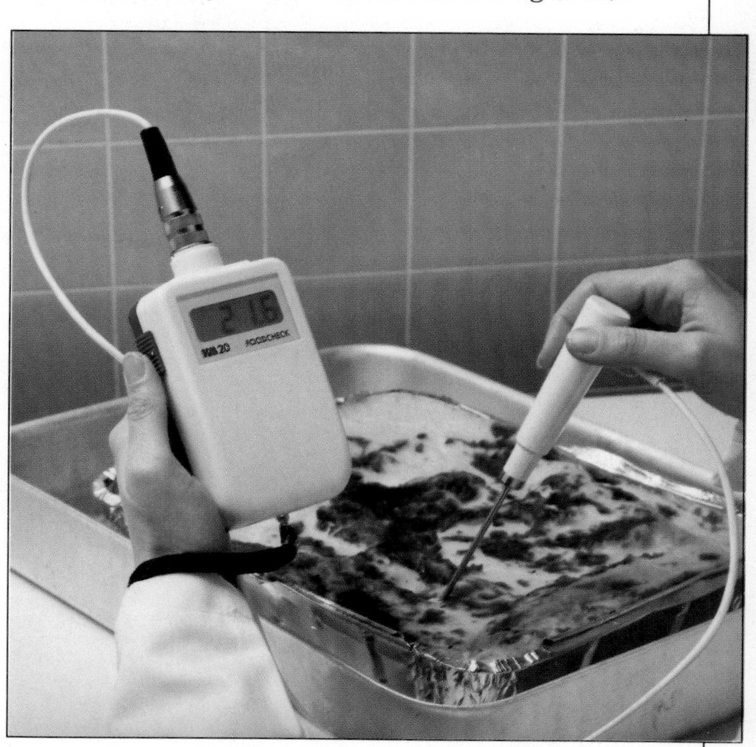

Make sure that 'heat and serve' food is heated correctly, according to instructions. Never cook it for less than the recommended time.

Food tasting

If you need to taste food during preparation, never use your fingers. Always use a clean spoon every time you taste the food, do not reuse.

Cooling food

Food must be cooled rapidly if it is not going to be eaten straight away. Never leave foods out for more than 1½ hours. You may need to split the food into portions to make sure that it cools in the time allowed.

Food should be cooled in an area away from other food preparation. The colder the area, the quicker the cooling. Don't put hot food straight into the fridge or cold store with other foods, as this raises the temperature in the fridge and causes condensation. Smaller items of food will cool quickest. Store cooled food in a refrigerator.

▲ Food must be cooled quickly and refrigerated within 1½ hours.

▲ Heat lamps keep food hot and a sneeze guard protects food from contamination by customers.

Displaying hot and cold food

Food displayed for sale needs to be protected against sources of contamination. All food should be covered so that it can't be contaminated by customers or staff.

Hot food must be kept at temperatures of above 63 °C by law.

Cold food must be stored at a safe temperature usually 8 °C or below. From 1 April 1993 the following cold foods must be kept below 5 °C:

- cut ripened soft cheeses
- cooked ready-to-eat foods which contain:
  - meat
  - fish
  - eggs
  - substitutes for meat, fish or eggs
  - cheese
  - cereals
  - pulses
  - vegetables
- smoked or cured fish
- slices of smoked or cured meats
- prepared sandwiches or rolls, containing meat, fish, eggs, soft cheeses or vegetables, (unless they are to be eaten on the same day as they are prepared).

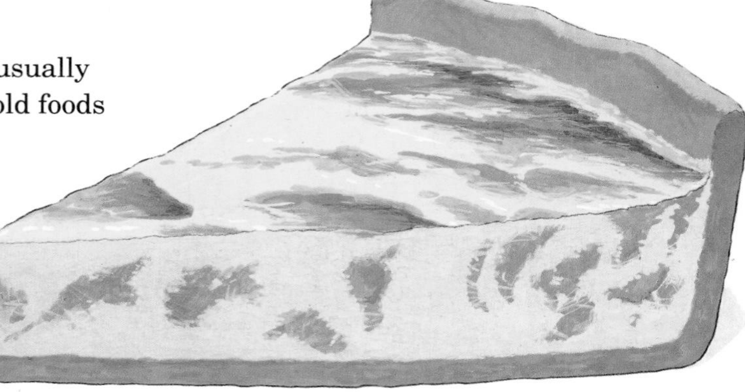

THE LAW

The temperature of specified foods must be 8 °C or below (for some foods this temperature must be reduced to 5 °C after 1 April 1993). Hot foods must be stored at 63 °C or above. These temperatures apply to food at all stages of the food chain, including distribution.

(Food Hygiene (Amendment) Regulations 1990/1991)

Avoiding contamination

Customers should not be able to handle unwrapped food before they buy it. If food is sold ready weighed, it should be wrapped.

Food handlers should use tongs or spoons to pick up food. There should be enough to make sure that the food can't be contaminated – one pair of tongs or spoon for each kind of food is best.

If you are using scales or a weighing machine, you should place the food on a piece of polythene or grease-proof paper or in a container. This stops any contamination of the scales.

Food should be covered or kept away from the serving counter, so that customers can't cough, sneeze or breathe over it. Counters should not be used for food storage or food preparation.

Never handle money and food at the same time. Money is a possible source of bacteria.

Animals (except guide dogs) are not allowed in food shops. Display signs on shop doors telling customers that this is not allowed.
Display 'No Smoking' signs in food shops.

SUMMARY

Planning food preparation can save time, energy and food.

Keep foods out of the danger zone (5 °C – 63 °C) for as long as possible.

Defrost food thoroughly in a cool area away from other foods. Do not refreeze.

Wash your hands after handling raw meat and poultry.

Use separate equipment when handling raw and cooked meats.

Take special care when dealing with high-risk foods.

Wash all equipment before and after using.

Make sure that all food is thoroughly cooked and hot all the way through before you serve it.

Use a clean spoon every time you taste food.

Food on display must be covered or kept away from a serving counter to avoid contamination from customers.

Hot food must be kept at a temperature of above 63 °C.

Cold food must be stored at a temperature of below 8 °C and in some cases below 5 °C, eg cooked meat.

Do not serve food with your hands. Tongs and spoons should be used.

Never put food directly on to weighing scales. Place it on grease-proof paper or polythene or in a container.

Never handle food and money at the same time.

Animals are not allowed in food shops.

Smoking is not allowed in food shops.

So now you know how to prepare food safely, but what about the places where food is prepared and sold? See chapter 9 ...

9

CHAPTER

Design of food premises

Well-designed food premises can help prevent food poisoning. Food premises should be easy to clean and built from suitable materials. All equipment used for food handling should be made from a suitable material, and should be used properly.

Food premises are anywhere food is stored, prepared or sold. This includes farms, vehicles used for delivering food, market stalls, food stands, food processing plants, shops, restaurants, ships and aircraft.

▲ These are all food premises.

INSIDE FOOD PREMISES

Food premises must be suitable for their use and they have to be registered with the local authority. There are very few which needn't be registered – for example, those that are open for less than five days in a period of five weeks.

THE LAW

Premises used for food business must be registered with their local authority. New businesses must register at least four weeks before opening.

(Food Premises (Registration) Regulations 1991)

Lighting and ventilation

Premises should be well lit and well ventilated. This makes an area safer, easier to clean and more comfortable to work in. The lighting should be good enough for you to see clearly what you are doing, with no shadows falling where you are working. The best ventilation is air-conditioning to avoid problems caused by dust or pests from open windows.

THE LAW

Food rooms must have satisfactory lighting and ventilation.

(Food Hygiene (General) Regulations 1970 as amended)

Storage

There should be enough space for safe storage of food. Food should be stored off the floor and there should be separate storage areas for raw foods and cooked foods. There must be enough refrigerated storage so that all food can be kept at a safe temperature.

Cleaning materials should be stored away from the food and off the floor. It is best if there is a separate area for cleaning equipment.

Staff facilities must have toilets, an area for storing outdoor clothing, and ideally, a rest area, where the staff can eat or smoke away from the food area.

Facilities

Food premises must have plenty of hot and cold water from a mains supply. There should be good drainage. The drains should be fitted with grease traps and any blockages should be cleared immediately. The premises should be pest free and there should be no easy access for food pests.

There must be separate facilities for hand washing and washing-up.

THE LAW

Wash-hand basins, with soap, nailbrush and drying facilities must be provided and kept clean. Hot water at a suitable temperature or hot and cold water must be supplied. There must be sufficient wash-hand basins so that they are convenient to use. Sinks for washing food or equipment must be provided and kept clean. Hot water must be supplied.

(Food Hygiene (General) Regulations 1970 as amended)

There should be a good supply of power points for electrical equipment. There should be no more than one plug to each socket.

There should be separate toilets for staff well away from the work area. All staff toilets must have a notice in them requesting users to wash their hands. A ventilated space must separate toilets from food rooms. Toilets should have wash basins, hot water, soap, a nail brush and hand-drying equipment. There should be toilets for customers in eating places.

There must be a well-equipped First Aid box on the premises. Everyone should know where to find it.

THE LAW

First Aid kits containing bandages and dressings, including waterproof dressings, must be provided in all food rooms.

(Food Hygiene (General) Regulations 1970 as amended)

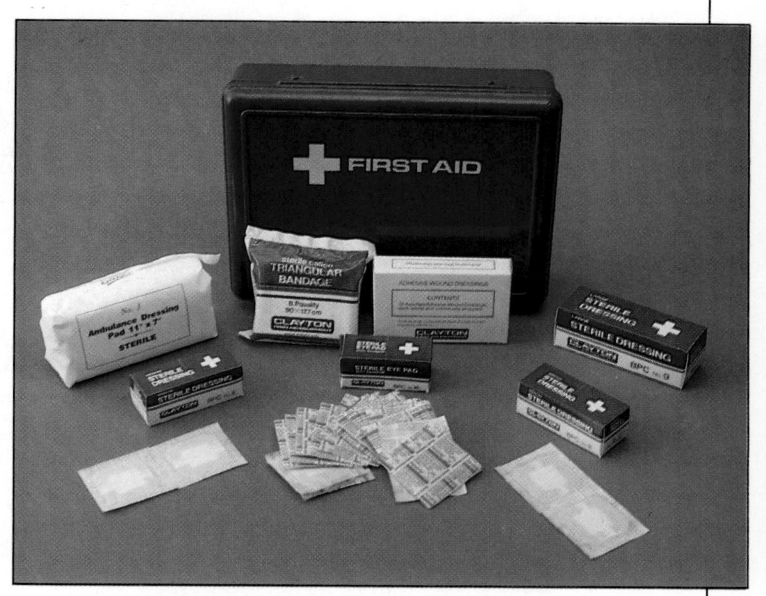

▲ All First Aid kits must contain these items.

There should be fire blankets and/or a suitable fire extinguisher. All staff should know where they are and how to use them.

Bins for kitchen refuse should have lids and be easy to clean. They should be placed in a clean area away from food preparation areas.

Layout

The premises should be pest free and there should be no easy access for food pests. Screens should be provided for open doors and windows.

There should be a well-thought-out work flow from delivery to the final product. This sequence should be from dirty to clean areas to avoid the risk of cross-contamination.

The area should be easy to clean. Places where ceilings and the walls or walls and floors meet should be rounded so no dirt can gather.

A food room must *not* be used as, or be next to, a sleeping place.

KEY WORDS

Food premises – areas where food is stored, transported, prepared or sold.

CONSTRUCTION

Walls, ceilings, floors, windows and woodwork must be in good order and condition so that they are easily cleaned and will keep pests away.

Walls and ceilings

The surfaces of walls and ceilings must be smooth. Painted surfaces should be smooth, not flaking. There should be no joints or cracks where dirt and bacteria can gather. Light colours are best so that any dirt will show up easily. Ceilings should be fire-resistant. Walls should be **impervious** (unable to absorb anything). Cavity walls sometimes provide hiding places for pests, so it is best to avoid them.

Floors

Floors should be made of a material which is easy to clean, but is non-slip. Tiled floors should be as even as possible so that dirt can't gather. The floor should have a slope so that any water on it can drain away easily and quickly.

Windows

If windows have to be opened for ventilation, they should have screens to stop pests getting in. If it is sunny, the areas near windows can get very hot, so food storage areas should never be near south-facing windows. Windows should be well fitting and kept clean. If they are wooden, the wood should be primed, then painted with a light-coloured gloss paint to give a shiny finish.

KEY WORDS

Impervious – unable to absorb anything. Waterproof.
Non-absorbent – water cannot pass through.
Heat-resistant – heat cannot pass through. Unlikely to be damaged by heat.

Work surfaces

Work surfaces should be smooth, with no cracks or joints, and made of a **non-absorbent**, easy-to-clean material. Surfaces should also be **heat-resistant**. Work surfaces should not be used as a store place and they should not be cluttered. Surfaces should *not* be made of wood, which is difficult to clean as it is absorbent.

Shelving

The best shelves are made of stainless steel. They should be easy to clean and cleaned regularly. All shelving should be safe and firmly fixed. It should be fastened to the wall but, if it is free-standing, it should not move when you lean on it. The bottom shelves must be off the floor.

EQUIPMENT

Fridges and freezers

There should be enough fridges and freezers so that all food which needs to be stored at safe temperatures can be put away. It is best if there are separate storage places for raw and cooked foods and for dairy foods. All fridges and freezers should be at the correct temperatures and away from direct sunlight.

Cookers

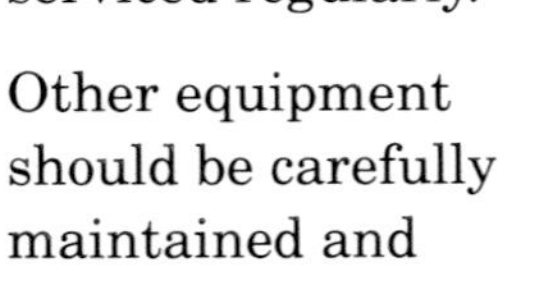

Cookers should be large enough to cook the amount of food being produced so that food doesn't have to be prepared too far in advance. They should be cleaned and serviced regularly.

Other equipment should be carefully maintained and cleaned. Machines like meat slicers should have safety guards and staff should be taught how to use them correctly.

Tableware

Crockery should be non-porous and dishwasher safe. Throw away any chipped or cracked crockery as it can harbour bacteria.

THE LAW

All equipment must be capable of being cleaned and be kept clean, and must be made of suitable materials. It must prevent any risk of contamination of the food.

(Food Hygiene (General) Regulations 1970 as amended)

Cutlery should be checked before it is used to make sure that it is clean. Pay special attention to areas between fork prongs and where the knife blades join the handles.

Glasses should not have smears. Polish them with a clean dry tea towel or with a disposable paper towel to get rid of water marks. Throw away any cracked or chipped glasses.

Report damaged equipment to your supervisor.

THE USE OF FOOD PREMISES

Careful planning can make food premises safer and more efficient.

There should be a flow of 'dirty' and 'clean' areas.

Storage areas should be near the delivery door. Deliveries should not be carried through food preparation areas.

Initial food preparation should be in an area near the stores, with a sink and wash-hand basin nearby.

Final food preparation can be in an area near cookers or the serving area.

Avoid unnecessary movement around a food area. One-way systems are useful in a crowded workplace as they prevent accidents if people are carrying hot food around.

If you have swing doors, use one to go in and one to go out. If this is not possible ensure that the door has a glass viewing panel. Try to clean as you go and always move food in to clean areas.

SUMMARY

Food premises should be well lit and well ventilated.

There should be plenty of hot and cold water, with good drainage.

There should be enough suitable storage space for foods, including cold storage (fridges, freezers) and enough cookers.

Food storage areas must be cool, dry and clean. Food should not be stored on the floor. There should be a separate area for cooling hot foods.

Pests should not be able to get into food premises.

Food premises must be easy to clean. Ceilings and walls should be smooth and floors non-slip.

Work surfaces should be made of a non-absorbent heat-resistant material which is easy to clean.

Wood is absorbent and so it should not be used in food premises.

Refuse should be kept in bins with lids well away from food preparation areas.

Toilets with wash basins must be provided. All staff toilets must have a notice stating 'Now wash your hands'. Staff should also have an area to store outdoor clothing, and ideally somewhere to eat and smoke.

To avoid cross-contamination, there should be separate food preparation areas for raw and cooked foods.

First Aid boxes must be provided and regularly maintained.

There should be enough equipment provided so that food can be prepared safely and efficiently.

Plan the use of the work areas so that there is a logical flow from delivery to storage to final food production. Try to clean as you go, with the food always moving to clean areas.

How do you keep food premises free from rubbish and food pests? Read chapter 10 ...

CHAPTER 10

Waste and pest control

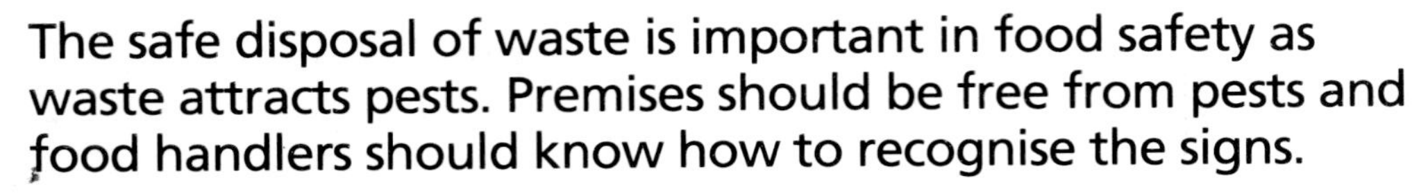

The safe disposal of waste is important in food safety as waste attracts pests. Premises should be free from pests and food handlers should know how to recognise the signs.

SAFE DISPOSAL OF WASTE

Inside the workplace

Waste food and scraps can be disposed of in a waste disposal unit. This is a machine which breaks down food waste and pipes it away. If your workplace doesn't have one, then put all waste into polythene bin liners inside bins with a well-fitting lid. Pedal bins are best, as you don't have to touch the bins with your hands. Bins should be in an area of the workplace which is well away from food preparation areas.

Always wrap sharp objects before putting them in a bin. They can cut people and split open the bin liners. Empty and clean bins regularly and always at the end of the day. Never let a bin get too full. Always wash your hands after handling rubbish.

Everything you want to recycle (eg aluminium and steel cans, glass bottles and jars) should be cleaned before storing. They should be kept away from food preparation areas. Arrange for them to be collected regularly.

▼ Pedal bins with lids are best.

Don't waste it!

RECYCLE

Dustbin bags left like this can encourage pests in to food premises. ▼

◀ Bins should be kept clean and tidy. They should have lids so that food pests cannot get in.

THE LAW

A suitable site for refuse storage must be provided for food premises. Refuse must not be allowed to build up in food rooms and around food premises

(Food Hygiene (General) Regulations 1970 as amended)

Refuse areas

Outside areas for waste storage should be away from the windows and doors of the workplace. The areas should be kept tidy and hosed down regularly. Bins or skips should have well-fitting lids so that they can't be blown off, and food pests cannot get in.

Bins should be emptied and cleaned out regularly. Get to know the collection times. If the bins are getting too full or aren't being emptied, tell your supervisor. You may need extra collections. Always wash your hands after taking refuse out to the bins.

PEST CONTROL

Food pests are any animals which live on our food. They contaminate foods by:

- eating the food and spreading bacteria from their saliva as they eat
- leaving droppings
- carrying bacteria on their bodies
- urinating on foods (rodents urinate often).

THE LAW

The structure of food rooms should be kept clean and in good repair to prevent infestations by rats, mice, insects or birds.

(Food Hygiene (General) Regulation 1970 as amended)

All food handlers should be able to recognise signs of food pests and know how to control them.

The best ways to stop pests are to make sure that they can't get in, and make sure that they can't get at the food.

Signs of pests will be much easier to see if the workplace is kept clean and tidy.

Stock rotation will help as it stops old stock being forgotten.

Check all new packages and containers for signs of pests and make sure that food is stored off the floor.

All loose foodstuff should be kept in airtight, pest-proof containers.

Rodents

The way in

Rats and mice can gnaw through wood and get in through gaps and holes left in walls or around doors. Make sure that there aren't any holes leading into the workplace from outside. Make sure that there are no gaps left where pipes enter the walls. Rats and mice can hide in undergrowth, so cut down all plants near the workplace.

▲ Rats and mice can carry diseases such as Salmonella on their fur and Weil's disease in their urine. They also do a lot of damage, by gnawing holes and making nests.

Signs

If you see, smell or hear rats or mice near food premises, you must tell your supervisor.
Obvious signs of infestations by rats or mice are droppings. Mouse droppings are small, black and oval in shape (about 1/4 cm long), rat droppings are larger (about 1 cm long). Rats also produce greasy smears around pipes. There may be worn runs (paths) in long grass which lead to the outside of a workplace. These will be caused by rats, they are timid and always tend to run along the same path.

Mice and rats gnaw at food packaging, and there may be tooth marks on pipes and on woodwork.
Rodents need to gnaw to keep their teeth short because they grow continuously. Look for holes in containers and food scattered around. If a mouse or rat has eaten through a container and been in contact with the food, throw the rest away immediately.

What to do

Rats and mice can be controlled either by trapping the animals, or by using poison. Rat poisons have to be used very carefully in food establishments. If your food premises has rats or mice you must tell your local authority who will give you advice.

KEY WORDS

Food pests – animals which live on and in our food.

Flying insects

The way in

Flying insects can get into food places through windows, doors and air bricks. You often find them around bins because they like food waste. Keep all rubbish in covered bins and put fly screens over air grills, and doors and windows which are often opened.

Signs

Adult flies can be seen easily.
Flies often lay eggs on food and food handlers should look out for signs of eggs and maggots in food. If you see these signs, don't use the food.

What to do

Flying insects can be controlled by using electric fly killers, or **pesticides**. Be careful to follow the instructions when using sprays – *never* spray near food. If there is a wasps' nest near the food place, contact your local authority for advice.

Crawling insects

Common food pests of this kind are cockroaches, book lice (psocids), ants and silverfish.

The way in

As these pests are so small, they get in through cracks, under doors and around pipes. Once inside buildings, they can live under units and in cracks. They live off food waste, so it is important to wipe up spills immediately and keep all loose foods in airtight containers.

Signs

Some insects (like cockroaches) are **nocturnal** and as they only come out to feed at night they can be difficult to spot. Cockroaches give off a sweetish smell. Insects like book lice and silverfish will be hard to see as they are small. Look out for dead bodies, egg-cases and droppings (black spots).

What to do

Most insects can be controlled using insecticides but be careful with these inside food preparation areas. *Never* spray insecticides near food. Tell your supervisor if you find signs of any insect pests.

Domestic pets and birds

Dogs, cats and wild birds can also spread bacteria to food. Never let pets inside food preparation areas.

Wild birds can carry Salmonella and Campylobacter (see page 11). Don't leave food out for the birds – it will encourage them to visit, and will also attract other food pests such as rats and mice.

Don't leave milk bottles where birds can peck at the tops.

KEY WORDS

Rodents – gnawing animals, eg rats and mice.
Pesticide – a chemical which kills pests. Insecticides kill insects, rodenticides kill rats and mice.
Nocturnal – animals which come out only at night, hiding away during the day.

SUMMARY

Dispose of food waste in a waste disposal unit or polythene bin liners kept inside a covered bin.

Clean things you want to recycle. Store them in a clean dry place off the floor.

Empty bins often. Never let them overflow.

Clean and disinfect bins regularly.

Outdoor bins and skips should be kept away from doors and windows of food preparation areas. They should have tight-fitting lids. Make sure that they are emptied regularly.

Keep outside areas tidy, and hose down regularly.

Food pests include rats, mice, flying insects and crawling insects. They spread bacteria and spoil food as they feed.

Signs of rodents are droppings and gnaw marks. Rats also leave greasy marks and make runs in long grass. Signs of insects are dead bodies, insect cases and droppings.

Make premises as pest free as possible. Fill all holes and cracks. Put fly-screens on windows, doors and air grills.

Make sure that there is no food lying about and that food is stored off the floor and in pest-proof containers.

Control pests with pesticides. If there are a lot of pests, contact your local authority for advice.

Wild birds carry disease – don't encourage them near a food area.

Don't let domestic pets come into food preparation areas.

What do you know?

Try these questions – the answers are in the book!

1 Put each of the following under the correct heading: Benefits of *good* food hygiene or Costs of *poor* food hygiene.

low staff turnover
satisfied customers
food poisoning
high staff turnover
legal action
food wastage
high profits
good appearance
infestations by pests.

BENEFITS	COSTS

2 What legal action can be taken if the Food Safety Act is not obeyed?

3 Bacteria need four things in order to multiply.

Fit them in this crossword puzzle.

4 Put these food contaminants under the correct headings:

False teeth	Bleach	Fly spray
Lead	Clostridium	Tin lid
Cherry stone	Copper	Staphylococcus.
Salmonella	Cigarette end	

PHYSICAL	CHEMICAL	BACTERIAL

5 Take a look around your local supermarket. Make a list of the high-risk foods on sale. What safety precautions does the shop take to prevent the food from becoming harmful?

6 List all the sources of bacteria in this kitchen.

7 Which of the drawings below shows

a direct contamination
b indirect contamination?

Try to identify the vehicle used by the bacteria.

8 On this picture of a thermometer, mark in the danger zone.

Match these sentences to the correct temperatures on the thermometer.

a Bacteria dormant (not multiplying)
b Bacteria multiplying
c Most bacteria destroyed

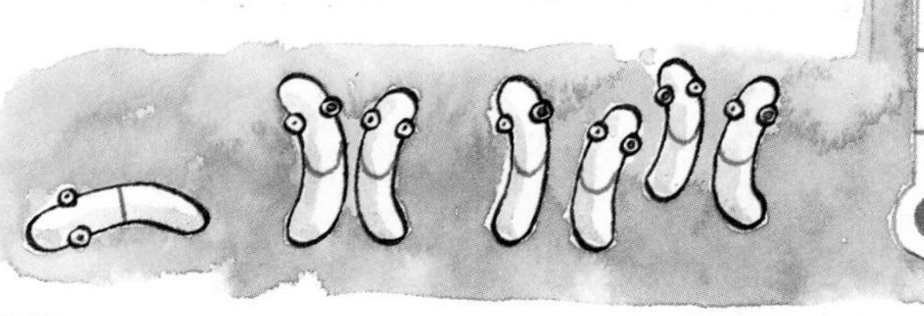

9 Explain how each of the methods of food preservation mentioned below keeps the food safe for longer.

Method	How it works
Sterilisation	Kills all bacteria present in food by heating to a high temperature for a long time
Pasteurisation	
Ultra heat treatment	

10 What is wrong in each of these pictures?

11 Put these in the correct order:

FINAL RINSE
DISINFECTION
DRY
PRE-CLEAN
RINSE
MAIN CLEAN

12 Match the words with their correct meanings:

DISINFECTANT
SANITISER
DETERGENT

- combined detergent and disinfectant
- chemical which dissolves grease
- chemical which reduces the number of bacteria to a safe level.

13 What is wrong with this food handler?

14 Look in your refrigerator at home. Use the tick list below to check the contents.

☐ All items are covered and in separate containers
☐ The fridge is at the correct temperature
☐ Cooked foods are kept above raw foods
☐ Raw meats are on the bottom shelf
☐ No food has passed its 'use by' date.

15 This is a plan for a kitchen. Mark on it where the best place is to:

a put a refrigerator
b store dried food
c store potatoes
d store canned food.

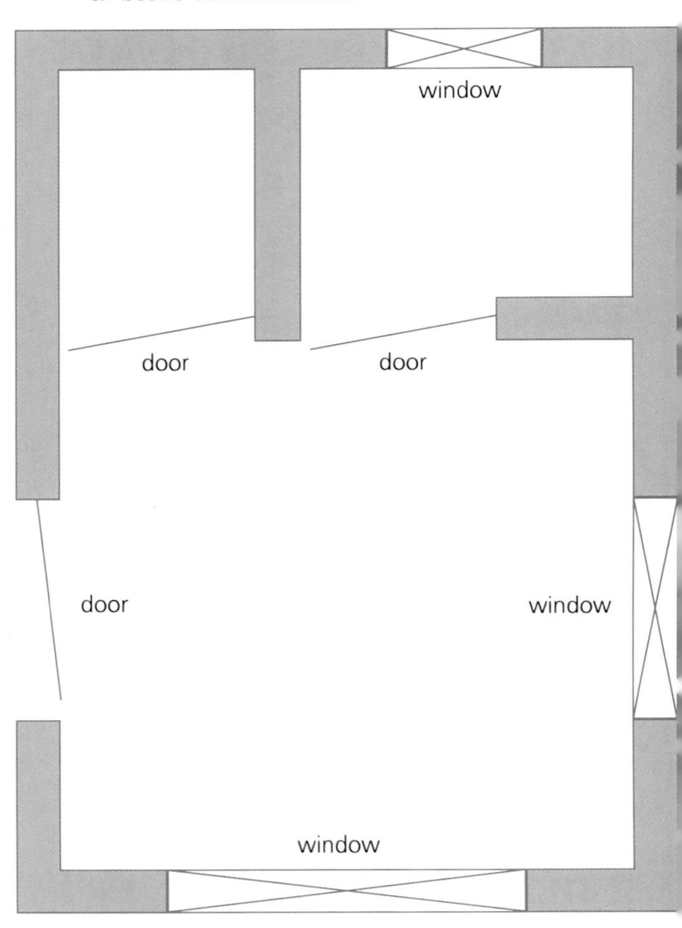

16 Match the correct temperatures with the correct storage:

room temperature ______
cool room ______
refrigerator ______
freezer ______

a –21 °C **b** 2 °C **c** 21 °C **d** 15 °C

17 Fill in the missing words in these sentences. Use the words from the list below:

clean spoon
63 °C
cooked
contamination
8 °C

Separate equipment should be used for ____________ and raw meats

Hot foods should be stored at a temperature above _____

The temperature of cold food should be below _____

Use a ________ ____________ each time when tasting food

If you thaw frozen food in the same area where cooked food is being cooled, you may get ______________________

18 Circle the examples of bad practice in this shop.

How would you put things right?

19 Choose the best word to complete these sentences:

1 In food premises a good material for a work surface is ____________________ WOOD / STAINLESS STEEL

2 In food premises ceilings should be painted a ___________ colour LIGHT / DARK

3 In food premises floors should be _________________ NON-SLIP / WAXED

4 In food premises fridges should be placed ________________ ________ AWAY FROM HEAT SOURCES/ ANYWHERE IN A KITCHEN

20 Spot at least three things wrong in this picture.

21 Fill in this table:

Food pests	Sign(s)	What to do
Rats and mice		
Flying insects		
Cockroaches		

If you don't know the answer to any of these questions, look back in the book, or ask your supervisor …

THE BASIC FOOD HYGIENE CERTIFICATE

The Basic Food Hygiene Certificate of the Institution of Environmental Health Officers (IEHO) is a first-level qualification in food hygiene and is nationally recognised in all sectors of the food industry.
The Certificate has been awarded to over 750,000 food handlers who have completed the course and assessment satisfactorily.

The course takes a minimum of six hours. The assessment may be of either the multi-choice tick-box type or an oral. Alternatively, a practical, competence-based route to assessment is available through some centres.

This coursebook has been designed as support material for candidates. Courses leading to the Certificate are offered at over 3,000 centres registered by the IEHO. There are centres throughout the United Kingdom and some overseas. These include:

- environmental health departments
- health authorities/hospitals
- colleges and schools
- public and private companies
- training organisations
- consultants

For details of local centres, contact the Training Department of the IEHO at the address below.

ENVIRONMENTAL HEALTH OFFICERS

Most Environmental Health Officers (EHOs) are employed by local authorities. An important part of EHOs' work is the enforcement of the food safety laws which exist to protect the public.

EHOs have the power to:

- enter, visit and inspect food premises at any reasonable time
- ensure that businesses are operating safely and hygienically
- identify unsafe food handling procedures
- close part or all of a business if necessary
- serve notices to improve conditions or stop certain practices
- prosecute employers and employees

THE INSTITUTION OF ENVIRONMENTAL HEALTH OFFICERS

The IEHO is a non-governmental organisation responsible for the training and professional development of over 8,000 EHOs working in England, Wales, Northern Ireland and other parts of the world. The Institution, which is a registered charity, was founded in 1883 and was granted a Royal Charter in 1984.

One of the main aims of the IEHO is to reduce the incidence of food poisoning.
Food poisoning is commonly caused by carelessness or ignorance.
Food handlers need training in food hygiene to ensure that they do not cause food poisoning.

The IEHO has established a leading role as a food hygiene Examining Body.
Its Basic Food Hygiene Certificate is designed to give food handlers essential knowledge and understanding of good food hygiene practices.

In addition to the Basic Food Hygiene Certificate, the Institution has designed two courses at higher levels. These lead to the Intermediate and Advanced Food Hygiene Certificates. For details of these courses, contact the Training Department at,
IEHO, 16 Great Guildford Street, London SE1 0ES.
Tel: 071 928 6006.

ISBN 0-900103-59-0
9 780900 103599